Lecture Notes in Mathematics 1854

Osamu Saeki

Topology of Singular Fibers of Differentiable Maps

Springer

Author

Osamu Saeki

Faculty of Mathematics
Kyushu University
Hakozaki, Fukuoka 812-8581, Japan
e-mail: saeki@math.kyushu-u.ac.jp

Library of Congress Control Number: 2004110903

Mathematics Subject Classification (2000): 57R45, 57N13

ISSN 0075-8434
ISBN 3-540-23021-1 Springer Berlin Heidelberg New York
DOI 10.1007/b100393

Springer is part of Springer Science+Business Media
springeronline.com

Typesetting: Camera-ready TEX output by the author

41/3142-543210 - Printed on acid-free paper

To Célia

Preface

In 1999, a friend of mine, Kazuhiro Sakuma, kindly asked me to give a series of lectures in the Kwansai Seminar on Differential Analysis, held at the Kinki University, Japan. At that time, I was studying the global topology of differentiable maps of 4-dimensional manifolds into lower dimensional manifolds. Sakuma and I had obtained a lot of interesting results concerning the relationship between the singularities of such maps and the differentiable structures of 4-dimensional manifolds; however, our results were not based on a systematic theory and were not satisfactory in a certain sense. So I was trying to construct such a systematic theory when I was asked to give lectures.

I wondered what kind of objects can reflect the *global* properties of manifolds. "Singularity" of a differentiable map can be such an object, but it is *local* in nature. I already knew that the notion of the Stein factorization played an important role in the global study of such maps; for example, refer to the works of Burlet–de Rham [7] or Kushner–Levine–Porto [28, 30]. Stein factorization is constructed by considering the connected components of the fibers of a given map.

This inspired me to consider singular fibers of differentiable maps. I promptly started the classification of singular fibers of stable maps of orientable 4-manifolds into 3-manifolds. It was not a difficult task, though quite tedious. Then I obtained the modulo two Euler characteristic formula in terms of the number of a certain singular fiber, by using Szűcs's formula [55], which Nuño Ballesteros and I had also obtained independently [36, 37]. The formula on the number of singular fibers was so beautiful that I was very happy to be able to present such a result in the Kwansai Seminar, in November 1999.

After attending my lectures, one of the participants, Toru Ohmoto, gave me a very important remark. He said "Your argument is closely related to Vassiliev's universal complex of multi-singularities. You just increased the number of generators for each cochain complex using the topology of singular fibers".

So I began to study Vassiliev's work and at the same time began to elaborate my results. It took a long time to write down all the details. A preprint version of the whole work was finished only in the middle of 2003, when I was staying in Strasbourg, France.

Now the acknowledgment follows. First of all, I would like to thank Kazuhiro Sakuma and the co-organizer Shuzo Izumi for kindly asking me to give a series of lectures at the Kwansai Seminar. I would like to thank Toru Ohmoto for his important remark at the seminar. Without these people, this work would have never appeared.

I would like to thank Minoru Yamamoto and Takahiro Yamamoto for carefully reading several earlier versions of the manuscript and for pointing out some important errors. I would also like to thank Goo Ishikawa, who gave me invaluable comments through his student Takahiro Yamamoto. I would also like to thank Jorge T. Hiratuka for stimulating discussions concerning Stein factorizations of stable maps of 4-dimensional manifolds.

I would like to express my sincere gratitude to András Szűcs for thoroughly reading an earlier version of the manuscript and for giving me many invaluable comments, which improved the manuscript considerably.

In January 2004, the results in this book were presented in a mini-course given at the University of Tokyo, Japan. I would like to thank all the participants at the mini-course, who attended it with enthusiasm and posed a lot of questions. In particular, I would like to thank Mikio Furuta for his excellent questions with fantastic ideas: in fact, I included some of the results based on his ideas in this book. I would also like to thank Masamichi Takase and Keiichi Suzuoka for their invaluable comments on my mini-course. I would like to thank Yukio Matsumoto, my ex-supervisor, for inviting me to give such a mini-course.

I would like to thank Vincent Blanlœil for inviting me to Strasbourg in 2003, where I could finish the first draft of this work. I would also like to express my thanks to Rustam Sadykov for posing many interesting questions concerning the book.

Finally, I would like to thank all the members of my family, especially to Célia, for their patience and support during the preparation of the book.

The author has been supported in part by Grant-in-Aid for Scientific Research (No. 16340018), Japan Society for the Promotion of Science.

Fukuoka,
July, 2004

Osamu Saeki

Contents

Introduction

Let $f : M \to N$ be a proper differentiable[1] map of an n-dimensional manifold M into a p-dimensional manifold N. When the codimension $p - n$ is nonnegative, for any point y in the target N, the inverse image $f^{-1}(y)$ consists of a finite number of points, provided that f is generic enough. Hence, in order to study the semi-local behavior of a generic map f around (the inverse image of) a point $y \in N$, we have only to consider the multi-germ $f : (M, f^{-1}(y)) \to (N, y)$. Therefore, we can use the well-developed theory of multi-jet spaces and their sections in order to study such semi-local behaviors of generic maps.

However, if the codimension $p - n$ is strictly negative, then the inverse image $f^{-1}(y)$ is no longer a discrete set. In general, $f^{-1}(y)$ forms a complex of positive dimension $n - p$. Hence, we have to study the map germ $f : (M, f^{-1}(y)) \to (N, y)$ along a set $f^{-1}(y)$ of positive dimension and the theory of multi-jet spaces is not sufficient any more. Surprisingly enough, there has been no systematic study of such map germs in the literature, as long as the author knows, although we can find some studies of the multi-germ of f at the singular points of f contained in $f^{-1}(y)$.

In this book, we consider the codimension -1 case, i.e. the case with $n - p = 1$, and classify the right-left equivalence classes of generic map germs $f : (M, f^{-1}(y)) \to (N, y)$ for $n = 2, 3, 4$. For the case $n = 3$, Kushner, Levine and Porto [28, 30] classified the singular fibers of C^∞ stable maps of 3-manifolds into surfaces up to *diffeomorphism*; however, they did not mention a classification up to *right-left equivalence* (for details, see Definition 1.1 (2) in Chap. 1). In this book, we clarify the difference between the classification up to diffeomorphism and that up to right-left equivalence by completely classifying the singular fibers up to these two equivalences.

Given a generic map $f : M \to N$ of negative codimension, the target manifold N is naturally stratified according to the right-left equivalence classes

[1]In this book, "differentiable" means "differentiable of class C^∞". We also use the term "smooth".

Fig. 0.1. The singular fiber whose number has the same parity as the Euler characteristic of the source 4-manifold M

of f-fibers. By carefully investigating how the strata are incident to each other, we get some information on the homology class represented by a set of the points in the target whose associated fibers are of certain types. This leads to some limitations on the co-existence of singular fibers. For example, we show that for a C^∞ stable map of a closed orientable 4-manifold into a 3-manifold, the number of singular fibers containing both a cusp point and a fold point is always even.

As an interesting and very important consequence of such co-existence results, we show that for a C^∞ stable map $f : M \to N$ of a closed orientable 4-manifold M into a 3-manifold N, the Euler characteristic of the source manifold M has the same parity as the number of singular fibers as depicted in Fig. 0.1 (Theorem 5.1). Note that this type of result would be impossible if we used the multi-germs of a given map at the singular points contained in a fiber instead of considering the topology of the fibers. In other words, our idea of essentially using the topology of singular fibers leads to new information on the global structure of generic maps.

Furthermore, the natural stratification of the target manifold according to the fibers enables us to generalize Vassiliev's universal complex of multi-singularities [58] to our case. In this book, we define such universal complexes of singular fibers and compute the corresponding cohomology groups in certain cases. It turns out that cohomology classes of such complexes give rise to cobordism invariants for maps with a given set of singularities in the sense of Rimányi and Szűcs [40].

The book is organized as follows.

In Part I, we define and study equivalence relations for singular fibers of generic differentiable maps and carry out the classification of singular fibers for some specific classes of maps. We use these classifications to obtain some results on the co-existence of singular fibers, and on the relationship between the numbers of certain singular fibers of stable maps and the topology of the source manifolds. We also give explicit concrete examples of such stable maps exhibiting typical singular fibers.

In Part II, we formalize the idea used to obtain the co-existence results of singular fibers in Part I in a more general setting. This leads to the notion of the universal complex of singular fibers, which is a refinement of Vassiliev's universal complex of multi-singularities. We develop a rather detailed theory of universal complex of singular fibers, and at the same time we give explicit calculations based on Part I. We will see that the cohomology classes of the universal complex of singular fibers give rise to invariants of cobordisms of singular maps in the sense of Rimányi and Szűcs [40] in the negative codimension case.

In Part III, we give some applications of our theory to the global topology of differentiable maps and present some further developments of the theory given in this book.

Part I consists of six chapters, which are organized as follows.

In Chap. 1, we give precise definitions of certain equivalence relations among the fibers of proper smooth maps, which will play essential roles in this book.

In Chap. 2, in order to clarify our idea, we classify the fibers of proper Morse functions on surfaces. The result itself should be folklore; however, we give a rather detailed argument, since similar arguments will be used in subsequent chapters.

In Chap. 3, we classify the fibers of proper C^∞ stable maps of orientable 4-manifolds into 3-manifolds up to right-left equivalence. Our strategy is to use a combinatorial argument, for obtaining all possible 1-dimensional complexes, together with a classification up to right equivalence of certain multi-germs due to [11, 61]. After the classification, we will see that the equivalence up to diffeomorphism and that up to right-left equivalence are almost equivalent to each other in our case. Furthermore, as another consequence of the classification, we will see that two fibers of such stable maps are C^0 right-left equivalent if and only of they are C^∞ right-left equivalent. This is an analogy of Damon's result [10] for C^∞ stable map germs in nice dimensions. Furthermore, we give similar results for proper C^∞ stable maps of (not necessarily orientable) 3-manifolds into surfaces and for proper C^∞ stable Morse functions on surfaces. For Morse functions on surfaces, we prove the following very important result: for two proper C^∞ stable Morse functions on surfaces, they are C^0 equivalent if and only if they are C^∞ equivalent.

In Chap. 4, we investigate the stratification of the target 3-manifold of a C^∞ stable map of a closed orientable 4-manifold as mentioned above and obtain certain relations among the numbers (modulo two) of certain singular fibers.

In Chap. 5, we combine the results obtained in Chap. 4 with the following two results. One is a result of Fukuda [14] and the author [45] about the Euler characteristics of the source manifold and the singular point set, and the other is Szűcs' formula [55] on the number of triple points of a generic surface in 3-space (see also [36, 37]). As a result, we obtain a congruence modulo two between the Euler characteristic of the source 4-manifold and the number of singular fibers as depicted in Fig. 0.1.

In Chap. 6, we construct explicit examples of C^∞ stable maps of closed orientable 4-manifolds into $\mathbf{R}^3$. Since $(4,3)$ is a nice dimension pair in the sense of Mather [32], given a 4-manifold M and a 3-manifold N, we have a plenty of C^∞ stable maps of M into N. However, it is surprisingly difficult to give an *explicit* example and to give a detailed description of the structure of the fibers. Here, we carry this out, and at the same time we explicitly construct infinitely many closed orientable 4-manifolds with odd Euler characteristics which admit smooth maps into $\mathbf{R}^3$ with only fold singularities. In the subsequent chapters, we will see that such explicit examples are essential and very important in the study of singular fibers of generic maps.

Part II consists of eight chapters as follows.

In Chap. 7, we generalize the idea given in Chaps. 4 and 5 in a more general setting to obtain certain results on the co-existence of singular fibers.

In Chap. 8, we define the universal complexes of singular fibers for proper Thom maps with coefficients in $\mathbf{Z}_2$, using an idea similar to Vassiliev's [58] (see also [23, 38]). Our universal complexes of singular fibers are very similar to Vassiliev's universal complexes of multi-singularities. In fact, we construct the complexes using the right-left equivalence classes of fibers instead of multi-singularities, and this corresponds to increasing the generators of each cochain group according to the topological structures of fibers. In order to use such universal complexes in several situations, we will develop a rather detailed theory of universal complexes of singular fibers. Here, given a set of generic maps and a certain equivalence relation among their fibers, we will define the corresponding universal complex of singular fibers.

In Chap. 9, we apply the general construction introduced in Chap. 8 to a more specific situation, namely in the case of proper C^∞ stable maps of orientable 4-manifolds into 3-manifolds. For such maps, we determine the structure of the universal complex of singular fibers with respect to a certain equivalence relation among the fibers and compute its cohomology groups explicitly.

In Chap. 10, we consider co-orientable fibers and construct the corresponding universal complex of co-orientable singular fibers with integer coefficients. We also give some important problems related to the theory of universal complexes of singular fibers.

In Chap. 11, we define a homomorphism induced by a generic map of the cohomology group of the universal complex of singular fibers to that of the target manifold of the map. This corresponds to associating to a cohomology class α of the universal complex the Poincaré dual to the homology class represented by the set of those points over which lies a fiber appearing in a cocycle representing α. We will see that the homomorphisms induced by explicit generic maps will be very useful in the study of the cohomology groups of the universal complexes. This justifies the study developed in Chap. 6.

In Chap. 12, we define a cobordism of smooth maps with a given set of singular fibers. We will see that the homomorphism defined in Chap. 11 restricted to a certain subgroup is an invariant of such a cobordism. Furthermore, we will give a criterion for a certain cochain of the universal complex of singular fibers to be a cocycle in terms of the theory of such cobordisms, and apply it to finding a certain nontrivial cohomology class of a universal complex associated to stable maps of 5-dimensional manifolds into 4-dimensional manifolds.

In Chap. 13, we consider cobordisms of smooth maps with a given set of local singularities in the sense of [40]. We explain how a cohomology class of a universal complex of singular fibers gives rise to a cobordism invariant for such maps. Note that such cobordism relations have been thoroughly studied in [40] in the nonnegative codimension case. Our idea provides a systematic and new method to construct cobordism invariants for negative codimension cases.

In Chap. 14, we give explicit examples of cobordism invariants constructed by using the method introduced in the previous chapters. In particular, we show that this method provides a complete invariant of fold cobordisms of Morse functions on closed oriented surfaces.

Part III consists of two chapters as follows.

In Chap. 15, we give explicit applications of the general idea given in Chap. 7 to the topology of certain generic differentiable maps. For example, we study the homology classes represented by some multiple point sets of certain generic maps. As a corollary, we show the vanishing of the Gysin image of a Stiefel-Whitney class for smooth maps under certain dimensional assumptions.

Finally in Chap. 16, we present some further results (without any details) concerning the topology of singular fibers of generic maps obtained after the first version of this book was written as a preprint.

Throughout this book, all manifolds and maps are differentiable of class C^∞. The symbol "$\cong$" denotes a diffeomorphism between manifolds or an appropriate isomorphism between algebraic objects. For a space X, the symbol "id_X" denotes the identity map of X. For other symbols used in this book, refer to the list starting at p. 135.

Part I

Classification of Singular Fibers

1

Preliminaries

In this chapter, we give some fundamental definitions, which will be essential for the classification of singular fibers of generic maps of negative codimensions.

Definition 1.1. (1) Let M_i be smooth manifolds and $A_i \subset M_i$ be subsets, $i = 0, 1$. A continuous map $g : A_0 \to A_1$ is said to be *smooth* if for every point $q \in A_0$, there exists a smooth map $\widetilde{g} : V \to M_1$ defined on a neighborhood V of q in M_0 such that $\widetilde{g}|_{V \cap A_0} = g|_{V \cap A_0}$. Furthermore, a smooth map $g : A_0 \to A_1$ is a *diffeomorphism* if it is a homeomorphism and its inverse is also smooth. When there exists a diffeomorphism between A_0 and A_1, we say that they are *diffeomorphic.*[1]

(2) Let $f_i : M_i \to N_i$ be smooth maps, $i = 0, 1$. For $y_i \in N_i$, we say that the fibers over y_0 and y_1 are *diffeomorphic* (or *homeomorphic*) if $(f_0)^{-1}(y_0) \subset M_0$ and $(f_1)^{-1}(y_1) \subset M_1$ are diffeomorphic in the above sense (resp. homeomorphic in the usual sense). Furthermore, we say that the fibers over y_0 and y_1 are C^∞ *equivalent* (or C^0 *equivalent*), if for some open neighborhoods U_i of y_i in N_i, there exist diffeomorphisms (resp. homeomorphisms) $\widetilde{\varphi} : (f_0)^{-1}(U_0) \to (f_1)^{-1}(U_1)$ and $\varphi : U_0 \to U_1$ with $\varphi(y_0) = y_1$ which make the following diagram commutative:

$$\begin{array}{ccc}
((f_0)^{-1}(U_0), (f_0)^{-1}(y_0)) & \xrightarrow{\widetilde{\varphi}} & ((f_1)^{-1}(U_1), (f_1)^{-1}(y_1)) \\
{\scriptstyle f_0}\downarrow & & \downarrow{\scriptstyle f_1} \\
(U_0, y_0) & \xrightarrow{\varphi} & (U_1, y_1).
\end{array} \tag{1.1}$$

When the fibers over y_0 and y_1 are C^∞ (or C^0) equivalent, we also say that the map germs $f_0 : (M_0, (f_0)^{-1}(y_0)) \to (N_0, y_0)$ and $f_1 : (M_1, (f_1)^{-1}(y_1)) \to (N_1, y_1)$ are smoothly (or topologically) *right-left equivalent.* Note that then

[1]Note that even if A_0 and A_1 are diffeomorphic to each other, the dimensions of the ambient manifolds M_0 and M_1 may be different.

$(f_0)^{-1}(y_0)$ and $(f_1)^{-1}(y_1)$ are diffeomorphic (resp. homeomorphic) to each other in the above sense.[2]

In what follows, if we just say "equivalent", or "right-left equivalent", then we mean "C^∞ equivalent" or "smoothly right-left equivalent", respectively.

When $y \in N$ is a regular value of a smooth map $f : M \to N$ between smooth manifolds, we call $f^{-1}(y)$ a *regular fiber*; otherwise, a *singular fiber*.

Example 1.2. If $f : M \to N$ is a proper submersion, then every fiber is regular. Furthermore, by Ehresmann's fibration theorem [13] (for details, see Theorem 1.4 below), the fibers over two points y_0 and $y_1 \in N$ are equivalent, provided that y_0 and y_1 belong to the same connected component of N. Thus each equivalence class corresponds to a union of connected components of N.

Example 1.3. Suppose that $f : M \to N$ is a Thom map, which is a stratified map with respect to Whitney regular stratifications of M and N such that it is a submersion on each stratum and satisfies a certain regularity condition (for more details, refer to [15, Chapter I, §3], [12, §2.5], [9, §2], [54], for example).

Let Σ be a stratum of N of codimension κ. Take a point $y \in \Sigma$ and let B_y be a small κ-dimensional open disk in N centered at y which intersects Σ transversely at the unique point y and is transverse to all the strata of N. Then by Thom's second isotopy lemma (for example, see [15, Chapter II, §5]), we see that the fiber of f over y is C^0 equivalent to the fiber of $(f|_{f^{-1}(B_y)}) \times \mathrm{id}_{\mathbf{R}^{p-\kappa}}$ over $y \times 0$, where $p = \dim N$. Thus, again by Thom's second isotopy lemma, we see that *the fibers over any two points belonging to the same stratum Σ of N are C^0 equivalent to each other.* Thus, each C^0 equivalence class corresponds to a union of strata of N.

Let us state Ehresmann's fibration theorem [13] in a form which will be useful in the subsequent chapters. For our purposes, we present here its relative version (see [29, §3]). Note that a continuous map is *proper* if the inverse image of a compact set is always compact.

Theorem 1.4. *Let $f : M \to \mathrm{Int}\, D^p$ be a proper submersion of an n-dimensional manifold M (possibly with boundary) into the interior of the p-dimensional disk with $n > p$ such that $f|_{\partial M} : \partial M \to \mathrm{Int}\, D^p$ is also a submersion. Then for the center 0 of $\mathrm{Int}\, D^p$, the inverse image $f^{-1}(0)$ is a compact $(n-p)$-dimensional manifold with boundary, and for an arbitrary diffeomorphism $h : f^{-1}(0) \to F$ onto a manifold F, there exists a diffeomorphism $\widetilde{h} : M \to F \times \mathrm{Int}\, D^p$ such that the diagram*

$$\begin{array}{ccc} M & \xrightarrow{\widetilde{h}} & F \times \mathrm{Int}\, D^p \\ & {}_{f}\searrow \quad \swarrow_{\pi} & \\ & \mathrm{Int}\, D^p & \end{array}$$

[2] Note that if the two fibers are equivalent in the above sense, then the dimensions of the source (or target) manifolds necessarily coincide with each other.

is commutative and that $\widetilde{h}|_{f^{-1}(0)} = h : f^{-1}(0) \to F \times \{0\}$, *where* $\pi : F \times \operatorname{Int} D^p \to \operatorname{Int} D^p$ *is the projection to the second factor.*

The above theorem can be proved as follows. We first construct a set of p vector fields on M projecting to the standard coordinate vector fields of $\operatorname{Int} D^p$, by using the partition of unity. Then by using the integral curves of the vector fields, we can construct the diffeomorphism $\widetilde{h}^{-1}$. (Here, we use the assumption that f is proper in order to guarantee the existence of required integral curves.) The diffeomorphism $h^{-1} : F \to f^{-1}(0)$ corresponds to the initial values of the integral curves. For details, see [6, §8.12], for example.

2

Singular Fibers of Morse Functions on Surfaces

Let us begin by the simplest case; namely, that of Morse functions on surfaces.

Let M be a smooth surface and $f : M \to \mathbf{R}$ a proper Morse function. For its critical points $c_1, c_2, \ldots \in M$, we assume that $f(c_i) \neq f(c_j)$ for $i \neq j$: i.e., we assume that each fiber of f contains at most one critical point. This is equivalent to saying that f is C^∞ stable (see, for example, [12, §4.3], [16, Chapter III, §2B]), so we often call such an f a *stable Morse function.*

By the Morse Lemma, at each critical point c_i, f is C^∞ right equivalent to the function germ of the form

$$(x, y) \mapsto \pm x^2 \pm y^2 + f(c_i)$$

at the origin. In particular, each singular fiber contains exactly one of the following two:

(1) a component consisting of just one point (corresponding to a local minimum or maximum),

(2) a "crossing point" which has a neighborhood diffeomorphic to

$$\{(x, y) \in \mathbf{R}^2 : x^2 - y^2 = 0,\, x^2 + y^2 < 1\}$$

(corresponding to a saddle point).

Since f is proper, each fiber of f is compact. Furthermore, for each regular point $q \in M$, the fiber through q is a regular 1-dimensional submanifold near the point. Hence the component of a singular fiber of f containing a critical point should be diffeomorphic to one of the three figures as depicted in Fig. 2.1 by a combinatorial reason.

More precisely, we can show the following.

Theorem 2.1. *Let $f : M \to \mathbf{R}$ be a proper stable Morse function on a surface M. Then the fiber over each critical value in $\mathbf{R}$ is equivalent to one of the three types of fibers as depicted in Fig. 2.2. Furthermore, two singular fibers of distinct types are not equivalent to each other even after taking the union with regular circle components.*

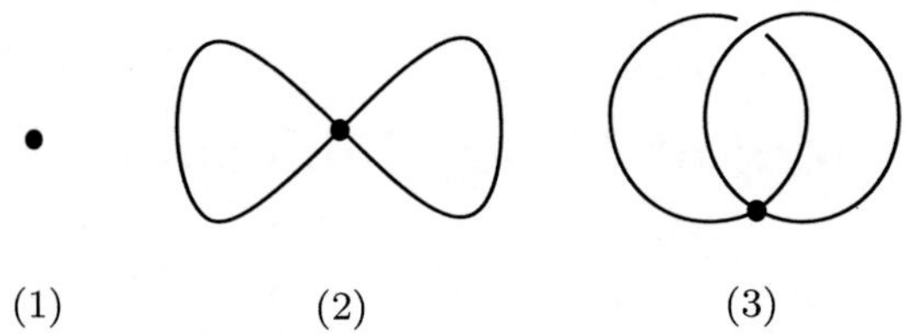

Fig. 2.1. List of diffeomorphism types of singular fibers for Morse functions on surfaces

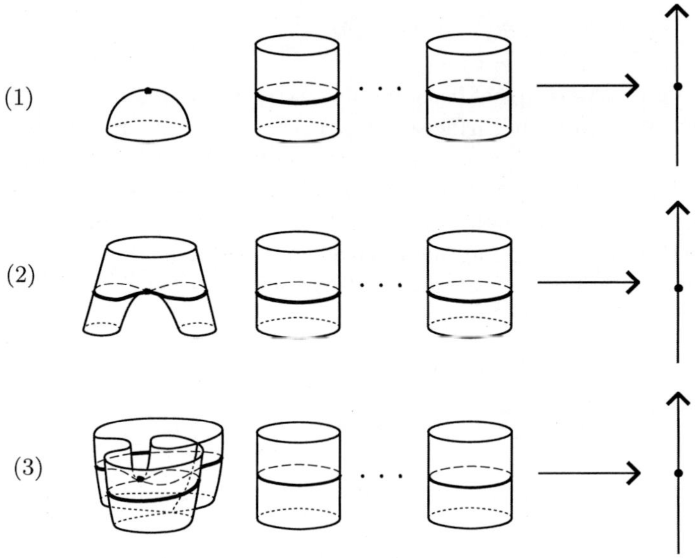

Fig. 2.2. List of equivalence classes of singular fibers for Morse functions on surfaces

Note that the source manifolds depicted in Fig. 2.2 are all open and have finitely many connected components. In particular, the source manifold of Fig. 2.2 (3) is diffeomorphic to the union of the once punctured open Möbius band and some copies of $S^1 \times \mathbf{R}$.

Proof of Theorem 2.1. If the corresponding critical point $c \in M$ is a local minimum or a local maximum, then the singular fiber is equivalent to that of Fig. 2.2 (1) by the Morse Lemma together with Ehresmann's fibration theorem [13] (see Theorem 1.4 in §1).

Suppose that c is a saddle point. By the Morse Lemma, the function germ of f at c is right equivalent to the function germ of $f_1 : (x, y) \mapsto x^2 - y^2$ at the origin up to a constant: i.e., there exists a diffeomorphism $\widetilde{\varphi}_1 : V \to V_1$ such that $\widetilde{\varphi}_1(c) = (0, 0)$ and $f_1 \circ \widetilde{\varphi}_1 = f - f(c)$ on V, where V is a neighborhood of c in M and V_1 is a neighborhood of the origin in $\mathbf{R}^2$ of the form

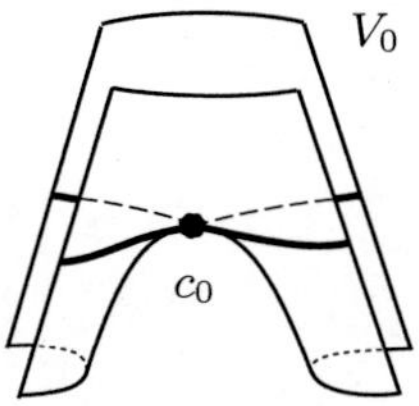

Fig. 2.3. The neighborhood V_0

$$V_1 = \{(x, y) \in \mathbf{R}^2 : x^2 + y^2 \leq \varepsilon, |f_1(x, y)| < \delta\}$$

for $1 >> \exists\varepsilon >> \exists\delta > 0$. In particular, there exists a diffeomorphism $\widetilde{\varphi}_0 : V \to V_0$ such that $\widetilde{\varphi}_0(c) = c_0$ and $f_0 \circ \widetilde{\varphi}_0 = f + (f_0(c_0) - f(c))$ on V, where f_0 is the Morse function as in Fig. 2.2 (2) or (3), which will be chosen later, c_0 is the critical point of f_0, and V_0 is the corresponding neighborhood of c_0 (see Fig. 2.3). Note that the maps

$$f|_{\partial V \cap f^{-1}((f(c)-\delta, f(c)+\delta))} : \partial V \cap f^{-1}((f(c) - \delta, f(c) + \delta)) \\ \to (f(c) - \delta, f(c) + \delta) \quad (2.1)$$

and

$$(f_0)|_{\partial V_0 \cap (f_0)^{-1}((f_0(c_0)-\delta, f_0(c_0)+\delta))} : \partial V_0 \cap (f_0)^{-1}((f_0(c_0) - \delta, f_0(c_0) + \delta)) \\ \to (f_0(c_0) - \delta, f_0(c_0) + \delta) \quad (2.2)$$

are proper submersions.

Since a Morse function is a submersion outside of the critical points, the closure of $f^{-1}(f(c)) \setminus V$ in M is a compact 1-dimensional smooth manifold whose boundary consists exactly of four points, and hence it is diffeomorphic to the disjoint union of two arcs and some circles. Therefore, $f^{-1}(f(c))$ is diffeomorphic to the disjoint union of (2) or (3) of Fig. 2.1 and some circles by a purely combinatorial reason. At this stage, we choose f_0 to be the Morse function as in Fig. 2.2 (2) (or (3)) if the component of $f^{-1}(f(c))$ containing c is diffeomorphic to (2) (resp. (3)) of Fig. 2.1. Furthermore, we choose the number of trivial circle bundle components appropriately.

When the component of $f^{-1}(f(c))$ containing c is diffeomorphic to (3) of Fig. 2.1, we see easily that the diffeomorphism

$$\widetilde{\varphi}_0|_{f^{-1}(f(c)) \cap V} : f^{-1}(f(c)) \cap V \to (f_0)^{-1}(f_0(c_0)) \cap V_0 \quad (2.3)$$

between the local fibers extends to a diffeomorphism between the whole fibers $f^{-1}(f(c))$ and $(f_0)^{-1}(f_0(c_0))$. In the case of Fig. 2.1 (2), this is not necessarily true (for example, see Fig. 2.4). If such an extension does not exist, then we modify the diffeomorphism $\widetilde{\varphi}_0$ by composing it with a self-diffeomorphism of V

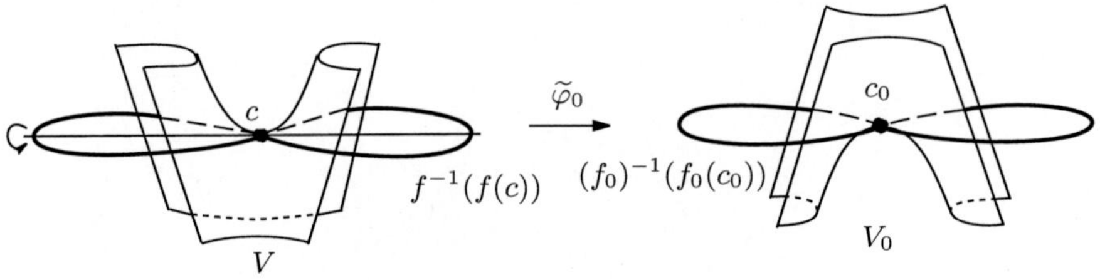

Fig. 2.4. A diffeomorphism $\widetilde{\varphi}_0 : V \to V_0$

corresponding to the diffeomorphism $h_1 : V_1 \to V_1$ defined by $(x, y) \mapsto (y, x)$ such that $f_1 \circ h_1 = -f_1$. Note that then we have $f_0 \circ \widetilde{\varphi}_0 = r \circ f$, where $r : \mathbf{R} \to \mathbf{R}$ is the reflection defined by $x \mapsto f_0(c_0) + f(c) - x$. Then we see that the diffeomorphism (2.3) between the local fibers extends to one between the whole fibers (see Fig. 2.4).

Since the maps (2.1) and (2.2) are proper submersions, we see that f (resp. f_0) restricted to $f^{-1}((f(c) - \delta, f(c) + \delta)) - \operatorname{Int} V$ (resp. $(f_0)^{-1}((f_0(c_0) - \delta, f_0(c_0) + \delta)) - \operatorname{Int} V_0$) is a smooth fibration over an open interval by virtue of the relative version of Ehresmann's fibration theorem (Theorem 1.4). Furthermore, the diffeomorphism $\widetilde{\varphi}_0 : V \to V_0$ can be extended to a fiber preserving diffeomorphism between $f^{-1}((f(c) - \delta, f(c) + \delta))$ and $(f_0)^{-1}((f_0(c_0) - \delta, f_0(c_0) + \delta))$. Hence we have the desired result.

The last statement in the theorem is clear. This completes the proof. □

Remark 2.2. Let $c \in M$ be a critical point of a proper stable Morse function $f : M \to \mathbf{R}$ on a surface M. Then for $\delta > 0$ sufficiently small, the difference

$$b_0(f^{-1}(f(c) + \delta)) - b_0(f^{-1}(f(c) - \delta))$$

is equal to ± 1 if c is of type (1) or (2), and is equal to 0 if c is of type (3), where b_0 denotes the 0-th betti number, or equivalently, the number of connected components.

Now let us examine the relationship among the numbers of singular fibers of the above three types. For a stable Morse function $f : M \to \mathbf{R}$ on a closed surface M, let $\mathbf{O}_{\text{odd}}$ denote the closure of the set

$$\{y \in \mathbf{R} : y \text{ is a regular value and } b_0(f^{-1}(y)) \text{ is odd}\}.$$

It is easy to see that $\mathbf{O}_{\text{odd}}$ is a finite disjoint union of closed intervals. Furthermore, a point $y \in \mathbf{R}$ is in $\partial\mathbf{O}_{\text{odd}}$ if and only if y is a critical value of type (1) or (2). Since the number of boundary points of a finite disjoint union of closed intervals is always even, we obtain the following.

Proposition 2.3. *Let $f : M \to \mathbf{R}$ be a stable Morse function on a closed surface M. Then the total number of singular fibers of types* (1) *and* (2) *is always even.*

Since the number of singular fibers is equal to the number of critical points, it has the same parity as the Euler characteristic $\chi(M)$ of the source surface M. Thus, we have the following.

Corollary 2.4. *Let $f : M \to \mathbf{R}$ be a stable Morse function on a closed surface M. Then the Euler characteristic $\chi(M)$ of M has the same parity as the number of singular fibers of type* (3).

Since a neighborhood of a singular fiber of type (3) is nonorientable, we immediately obtain the following special case of the Poincaré duality, using the fact that every closed surface admits a stable Morse function.

Corollary 2.5. *Every orientable closed surface has even Euler characteristic.*

By analyzing the singular fibers of type (3), we can also give an interesting proof for the following well-known fact.[1]

Proposition 2.6. *For a closed surface M, we always have*

$$w_1(M)^2 = w_2(M) \in H^2(M; \mathbf{Z}_2),$$

where $w_i(M)$ denotes the i-th Stiefel Whitney class of M.

Proof. Without loss of generality, we may assume that M is connected.

Let $f : M \to \mathbf{R}$ be an arbitrary stable Morse function on the surface M. Recall that a singular fiber of type (3) is a union of two nonsingular circles which intersect each other transversely, and that the tubular neighborhood of each of the two circles is a Möbius band. Note that if we cut the regular neighborhood of such a singular fiber along one of the two circles, then we get an orientable surface.

Let us take one of the two circles from each singular fiber of type (3) of f, and let C be their union. Let us denote by C' the union of the complementary circles. The surfaces $M \setminus C$ or $M \setminus C'$ may still be nonorientable. However, we can take a union $\widetilde{C}$ (or $\widetilde{C'}$) of some components of regular fibers of f so that $M \setminus (C \cup \widetilde{C})$ (resp. $M \setminus (C' \cup \widetilde{C'})$) is orientable. We may assume that $\widetilde{C}$ (or $\widetilde{C'}$) has the minimal number of circle components with this property. Then $C \cup \widetilde{C}$ (or $C' \cup \widetilde{C'}$) represents a homology class in $H_1(M; \mathbf{Z}_2)$ Poincaré dual to $w_1(M) \in H^1(M; \mathbf{Z}_2)$.

Let $[M] \in H_2(M; \mathbf{Z}_2)$ denote the fundamental class of the surface M. Then the number $\langle w_1(M)^2, [M]\rangle \in \mathbf{Z}_2$ is equal to the modulo 2 intersection number of $C \cup \widetilde{C}$ and $C' \cup \widetilde{C'}$ in M. By construction, we see easily that this is equal

[1]The author is indebted to Mikio Furuta for the idea of the proof.

to the number of singular fibers of type (3) modulo 2. By Corollary 2.4 this coincides with the Euler characteristic of M modulo 2, and hence with the number $\langle w_2(M), [M]\rangle \in \mathbf{Z}_2$. Since M is connected, this implies that $w_1(M)^2 = w_2(M)$. This completes the proof. □

Note that in the above proof, we did not use the classification theorem of closed surfaces.

Remark 2.7. Let M be a closed connected nonorientable surface of nonorientable genus g: i.e., M is homeomorphic to the connected sum of g copies of the real projective plane $\mathbf{R}P^2$. Then the number of singular fibers of type (3) of a stable Morse function on M is always less than or equal to g, since M can contain at most g disjointly embedded Möbius bands.

In fact, we have the following.[2]

Proposition 2.8. *Let M be a closed connected nonorientable surface of nonorientable genus g. Then for every*

$$k \in \{n \in \mathbf{Z} : 0 \leq n \leq g \text{ and } n \equiv \chi(M)(= 2 - g) \pmod 2\},$$

there exists a stable Morse function $f : M \to \mathbf{R}$ which has exactly k singular fibers of type (3).

Proof. It is easy to construct a Morse function $f_1 : \mathbf{R}P^2 \to \mathbf{R}$ on the real projective plane with exactly three critical points. Then the singular fiber passing through the critical point of index 1 is of type (3) and the other singular fibers are of type (1). Furthermore, let $f_0' : T^2 \to \mathbf{R}$ be the standard height function on the 2-dimensional torus. It is a stable Morse function with exactly four critical points, whose indices are equal to 0, 1, 1 and 2. Let c be a component of the regular fiber over a value between the two values of the critical points of index 1. Cutting T^2 along c and pasting the two circle boundaries by an orientation reversing diffeomorphism, we obtain a Klein bottle K^2 and a stable Morse function $f_0 : K^2 \to \mathbf{R}$. By construction, f_0 has no singular fibers of type (3).

For a given integer k as in the proposition, set $\ell = (g - k)/2$. Then by taking the "connected sum" of k copies of $f_1 : \mathbf{R}P^2 \to \mathbf{R}$ and ℓ copies of $f_0 : K^2 \to \mathbf{R}$, we obtain a stable Morse function on M with exactly k singular fibers of type (3). (Here, before performing a "connected sum" of two Morse functions, we add a constant to one of the functions so that the minimum of a function is greater than the maximum of the other function. For details, see [46].) This completes the proof. □

Remark 2.9. All the results in this chapter are valid also for maps into circles.

---

[2] This is an answer to a question of András Szűcs. The author would like to thank him for such an interesting question.

We end this chapter by an exercise, which can be solved by almost the same argument as in this chapter. The author hopes that the reader will enjoy solving it!

Exercise 2.10. Classify the singular fibers of proper stable Morse functions on 3-dimensional manifolds.

3

Classification of Singular Fibers

In this chapter, we consider proper C^∞ stable maps of orientable 4-manifolds into 3-manifolds, and classify their singular fibers up to the equivalences described in Definition 1.1. As a consequence, we see that two such fibers are C^0 equivalent if and only if they are C^∞ equivalent. We also study singular fibers of C^∞ stable maps of surfaces and 3-manifolds and show that two stable Morse functions on a surface are C^0 right-left equivalent if and only if they are C^∞ right-left equivalent.

3.1 Stable Maps of 4-Manifolds into 3-Manifolds

Let M be a 4-manifold and N a 3-manifold. The following characterization of C^∞ stable maps $M \to N$ is well-known.

Proposition 3.1. *A proper smooth map $f : M \to N$ of a 4-manifold M into a 3-manifold N is C^∞ stable if and only if the following conditions are satisfied.*

(i) *For every $q \in M$, there exist local coordinates (x, y, z, w) and (X, Y, Z) around $q \in M$ and $f(q) \in N$ respectively such that one of the following holds:*

$$(X \circ f, Y \circ f, Z \circ f) = \begin{cases} (x, y, z), & q\text{: } \textit{regular point}, \\ (x, y, z^2 + w^2), & q\text{: } \textit{definite fold point}, \\ (x, y, z^2 - w^2), & q\text{: } \textit{indefinite fold point}, \\ (x, y, z^3 + xz - w^2), & q\text{: } \textit{cusp point}, \\ (x, y, z^4 + xz^2 + yz + w^2), & q\text{: } \textit{definite swallowtail}, \\ (x, y, z^4 + xz^2 + yz - w^2), & q\text{: } \textit{indefinite swallowtail}. \end{cases}$$

(ii) *Set $S(f) = \{q \in M : \operatorname{rank} df_q < 3\}$, which is a regular closed 2-dimensional submanifold of M under the above condition* (i)*. Then, for*

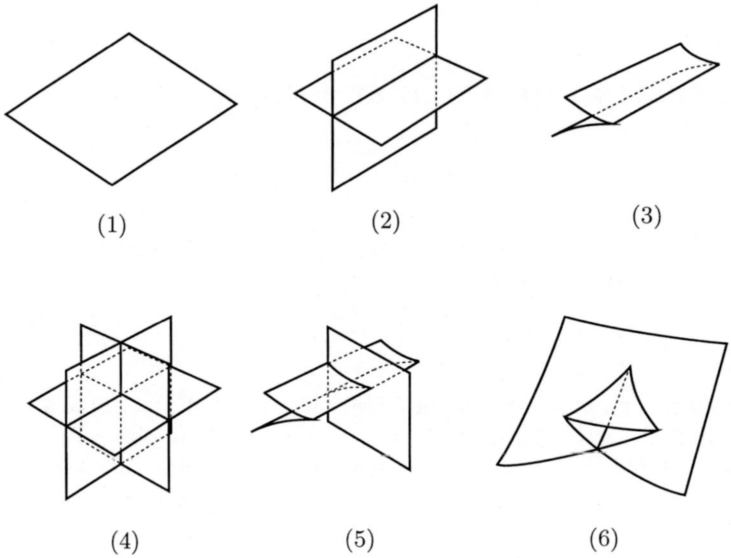

Fig. 3.1. Multi-germs of $f|_{S(f)}$

every $r \in f(S(f))$, $f^{-1}(r) \cap S(f)$ *consists of at most three points and the multi-germ*

$$(f|_{S(f)}, f^{-1}(r) \cap S(f))$$

is right-left equivalent to one of the six multi-germs as described in Fig. 3.1*:* (1) *represents a single immersion germ which corresponds to a fold point,* (2) *and* (4) *represent normal crossings of two and three immersion germs, respectively, each of which corresponds to a fold point,* (3) *corresponds to a cusp point,* (5) *represents a transverse crossing of a cuspidal edge as in* (3) *and an immersion germ corresponding to a fold point, and* (6) *corresponds to a swallowtail.*

Remark 3.2. According to du Plessis and Wall [12, 60], if (n, p) is in the nice range in the sense of Mather [32], a proper smooth map between manifolds of dimensions n and p is C^∞ stable if and only if it is C^0 stable. Hence, the above proposition gives a characterization of C^0 stable maps of 4-manifolds into 3-manifolds as well, since $(4, 3)$ is in the nice range.

Let q be a singular point of a proper C^∞ stable map $f : M \to N$ of a 4-manifold M into a 3-manifold N. Then, using the above local normal forms, it is easy to describe the diffeomorphism type of a neighborhood of q in $f^{-1}(f(q))$. More precisely, we easily get the following local characterizations of singular fibers.

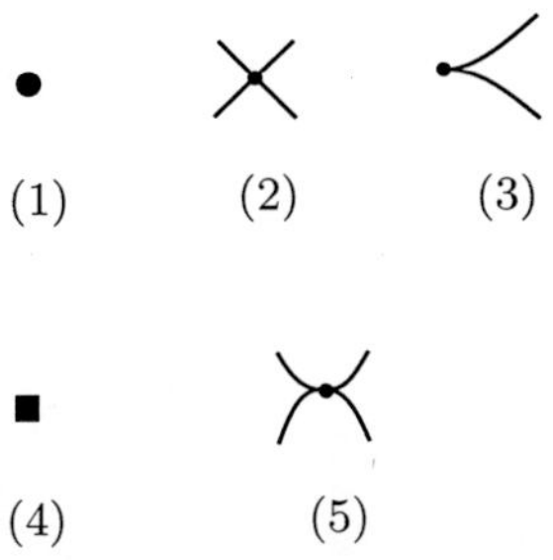

Fig. 3.2. Neighborhood of a singular point in a singular fiber

Lemma 3.3. *Every singular point q of a proper C^∞ stable map $f : M \to N$ of a 4-manifold M into a 3-manifold N has one of the following neighborhoods in its corresponding singular fiber (see Fig. 3.2):*

(1) *isolated point diffeomorphic to $\{(x,y) \in \mathbf{R}^2 : x^2 + y^2 = 0\}$, if q is a definite fold point,*
(2) *union of two transverse arcs diffeomorphic to $\{(x,y) \in \mathbf{R}^2 : x^2 - y^2 = 0\}$, if q is an indefinite fold point,*
(3) *cuspidal arc diffeomorphic to $\{(x,y) \in \mathbf{R}^2 : x^3 - y^2 = 0\}$, if q is a cusp point,*
(4) *isolated point diffeomorphic to $\{(x,y) \in \mathbf{R}^2 : x^4 + y^2 = 0\}$, if q is a definite swallowtail,*
(5) *union of two tangent arcs diffeomorphic to $\{(x,y) \in \mathbf{R}^2 : x^4 - y^2 = 0\}$, if q is an indefinite swallowtail.*

Note that in Fig. 3.2, both the black dot (1) and the black square (4) represent an isolated point; however, we use distinct symbols in order to distinguish them.

For the local nearby fibers, we have the following.

Lemma 3.4. *Let $f : M \to N$ be a proper C^∞ stable map of a 4-manifold M into a 3-manifold N and $q \in S(f)$ a singular point such that $f^{-1}(f(q)) \cap S(f) = \{q\}$. Then the local fibers near q are as in Fig. 3.3:*

(1) *q is a definite fold point,*
(2) *q is an indefinite fold point,*
(3) *q is a cusp point,*
(4) *q is a definite swallowtail,*
(5) *q is an indefinite swallowtail,*

where each 0- or 1-dimensional object represents a portion of the fiber over the corresponding point in the target and each 2-dimensional object represents $f(S(f)) \subset N$ near $f(q)$.

In the following, we assume that the 4-manifold M is orientable. Then we get the following classification of singular fibers.

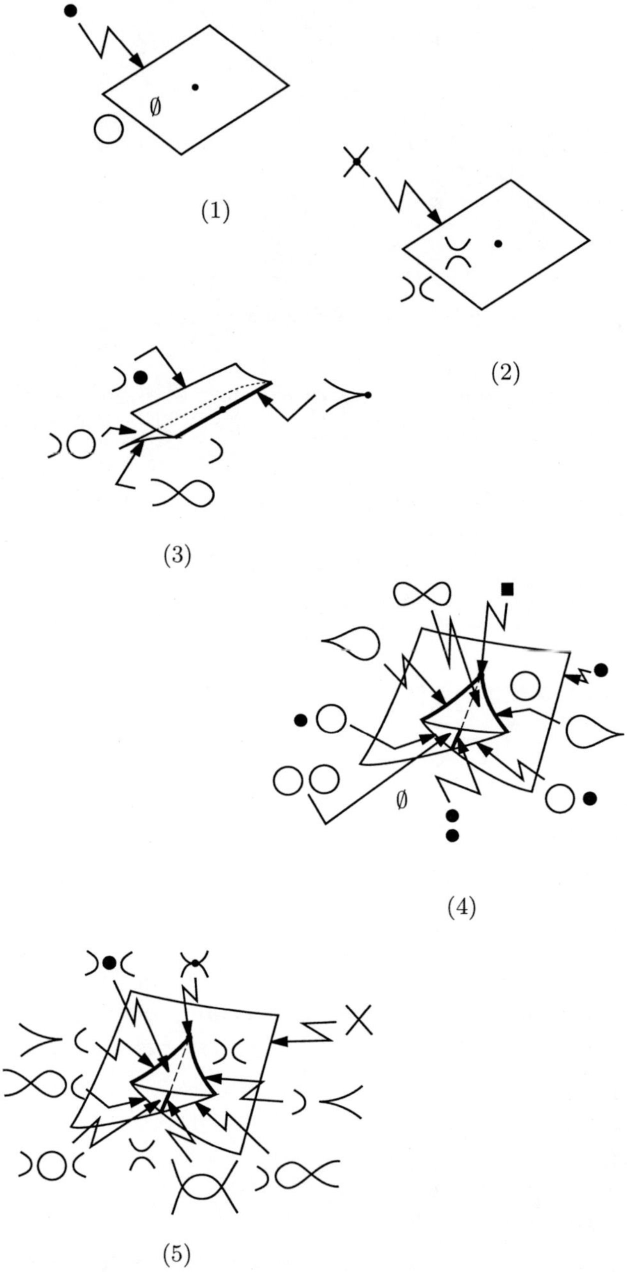

Fig. 3.3. Local degenerations of fibers

Theorem 3.5. *Let $f : M \to N$ be a proper C^∞ stable map of an orientable 4-manifold M into a 3-manifold N. Then, every singular fiber of f is equivalent to the disjoint union of one of the fibers as in Fig. 3.4 and a finite number of copies of a fiber of the trivial circle bundle. Furthermore, no two fibers appearing in the list are equivalent to each other even after taking the union with regular circle components.*

In Fig. 3.4, κ denotes the codimension of the set of points in N whose corresponding fibers are equivalent to the relevant one. For details, see Remark 3.7. Furthermore, $\mathrm{I}^*, \mathrm{II}^*$ and III^* mean the names of the corresponding singular fibers, and "/" is used only for separating the figures. Note that we have named the fibers so that each connected fiber has its own digit or letter, and a disconnected fiber has the name consisting of the digits or letters of its connected components. Hence, the number of digits or letters in the superscript coincides with the number of connected components.

It is not difficult to describe the behavior of the map on a neighborhood of each singular fiber in Fig. 3.4. This can also be regarded as a degeneration of fibers around the singular fiber, or a deformation of the singular fiber. In Figs. 3.5–3.8 are depicted the nearby fibers for four of the 27 singular fibers (Fig. 3.3 (1) and (4) can also be regarded as the deformations of the singular fibers of types I^0 and III^c respectively).[1] Since we are assuming that the source 4-manifold is orientable, the singular fiber as in Fig. 2.1 (3) never appears in the degenerations.

Remark 3.6. Each singular fiber described in Fig. 3.4 can be realized as a component (or as a union of some components) of a singular fiber of some C^∞ stable map of a *closed* orientable 4-manifold into $\mathbf{R}^3$. This can be seen as follows. Given a singular fiber, we can first realize it semi-locally; i.e., we can construct a proper C^∞ stable map of an open 4-manifold M_0 into $\mathbf{R}^3$ such that its image coincides with the open unit disk in $\mathbf{R}^3$ and that it has the given singular fiber over the center. Such a map can be constructed, for example, by using a 2-parameter deformation of smooth functions on an orientable surface: in this case, the open 4-manifold M_0 is diffeomorphic to the product of an open orientable surface and an open 2-disk (for example, refer to the construction in Chap. 6 using Fig. 6.3). Then we can extend the map to a smooth map of a closed orientable 4-manifold M containing M_0 into $\mathbf{R}^3$. Perturbing the extended map slightly, we obtain a desired stable map. In fact, we can choose an arbitrary closed orientable 4-manifold as the source manifold M of the desired map.

Proof of Theorem 3.5. Let us take a point $y \in f(S(f))$. We will first show that the union of the components of $f^{-1}(y)$ containing singular points is diffeomorphic to one of the fibers listed in Fig. 3.4 in the sense of Definition 1.1 (2).

[1] The degenerations of fibers around all the singular fibers are described in detail by colorful and beautiful figures in [20].

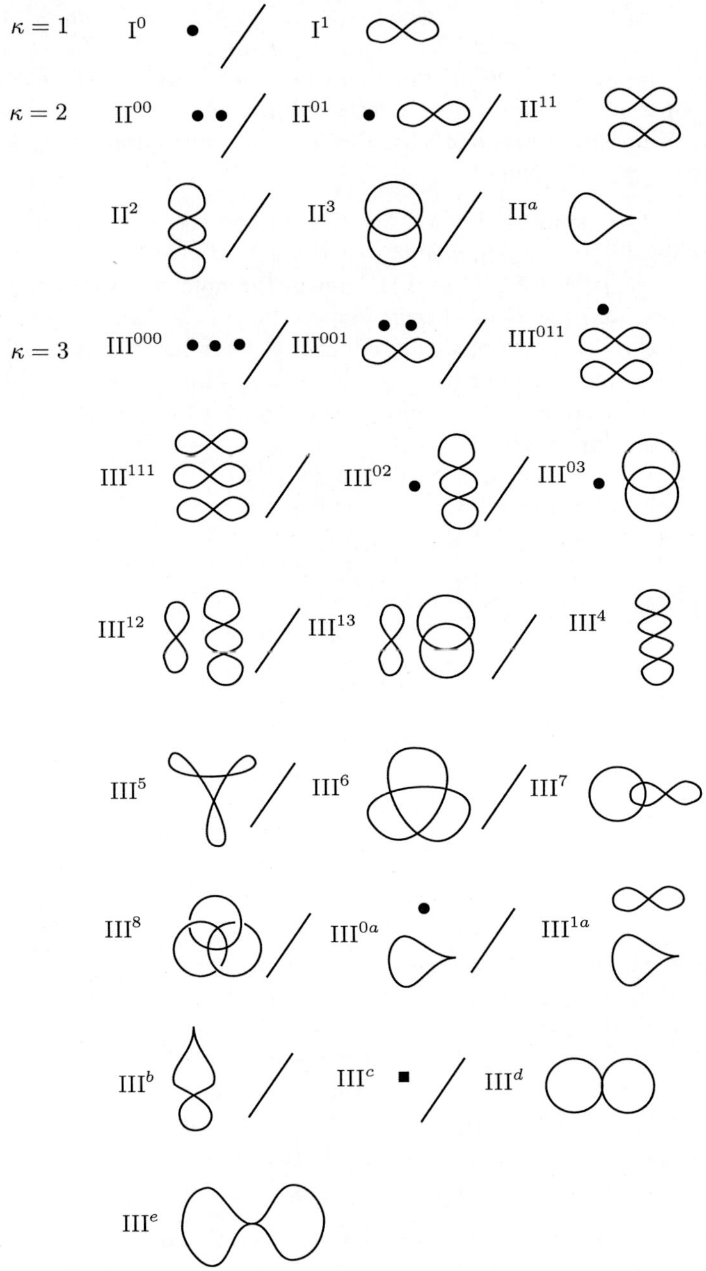

Fig. 3.4. List of singular fibers of proper C^∞ stable maps of orientable 4-manifolds into 3-manifolds

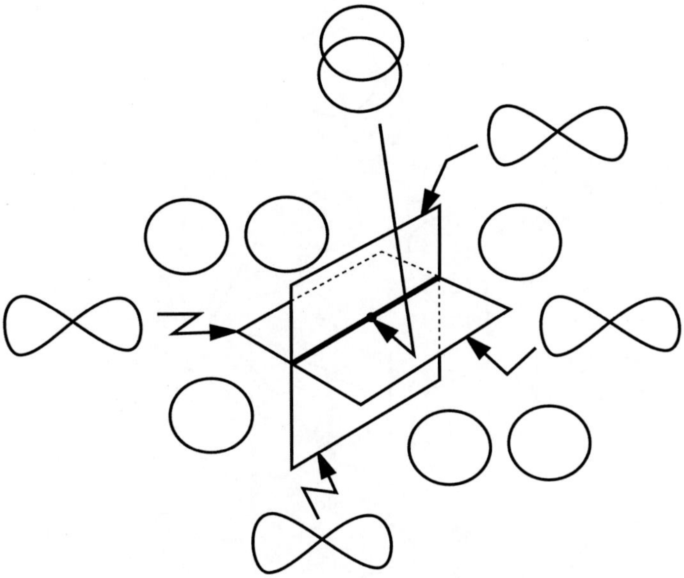

Fig. 3.5. Degeneration of fibers around the fiber of type II^3

If y corresponds to Fig. 3.1 (1), then $f^{-1}(y)$ contains exactly one singular point, which is a fold point. Thus, by an argument similar to that in the proof of Theorem 2.1, we see that the component of $f^{-1}(y)$ containing the singular point is diffeomorphic to one of the three figures of Fig. 2.1. If a fiber as in Fig. 2.1 (3) appears, then the 4-manifold M must contain a punctured Möbius band times D^2, and hence is nonorientable. Since we have assumed that M is orientable, this does not occur. Hence, we see that the singular fiber $f^{-1}(y)$ is diffeomorphic to the disjoint union of I^0 (or I^1) and a finite number of nonsingular circles.

If y corresponds to Fig. 3.1 (2), then $f^{-1}(y)$ contains exactly two singular points, say q_1 and q_2, which are fold points. Since they have neighborhoods as in Lemma 3.3 (1) or (2) in $f^{-1}(y)$, and since f is a submersion outside of the singular points, we see that there are only a finite number of possibilities for the diffeomorphism type of the union of the components of $f^{-1}(y)$ containing q_1 and q_2: for example, if both q_1 and q_2 are indefinite fold points, then it is obtained from two copies of the figure as in Fig. 3.2 (2) by connecting their end points by four arcs. Then we can use Lemma 3.4 to obtain the nearby fibers of each possible singular fiber: for example, for the singular fiber of type II^3, see Fig. 3.5. Excluding the possibilities such that a singular fiber as in Fig. 2.1 (3) appears as a nearby fiber, we get the fibers $\mathrm{II}^{00}, \mathrm{II}^{01}, \mathrm{II}^{11}, \mathrm{II}^2$ and II^3.

By similar combinatorial arguments, we obtain the following singular fibers:

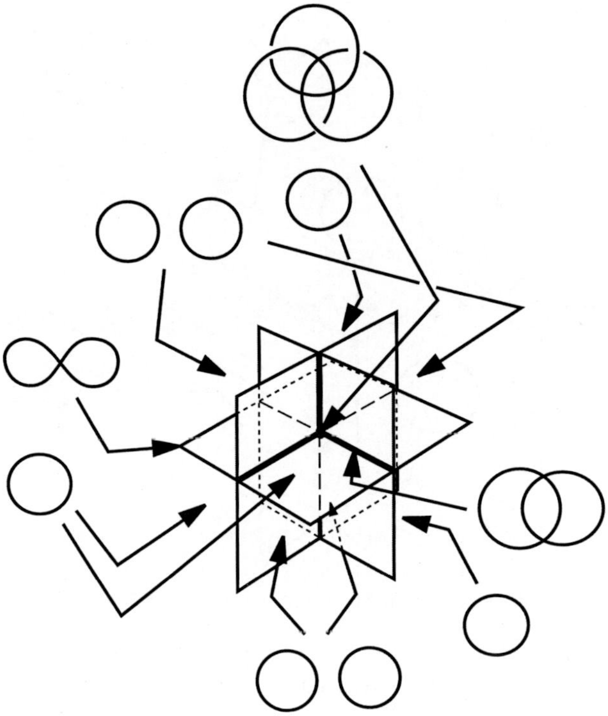

Fig. 3.6. Degeneration of fibers around the fiber of type III^8

(1) if y corresponds to Fig. 3.1 (3), then we obtain II^a,

(2) if y corresponds to Fig. 3.1 (4), then we obtain III^{000}, III^{001}, III^{011}, III^{111}, III^{02}, III^{03}, III^{12}, III^{13}, III^4, III^5, III^6, III^7 and III^8,

(3) if y corresponds to Fig. 3.1 (5), then we obtain $\mathrm{III}^{0a}, \mathrm{III}^{1a}$ and III^b,

(4) if y corresponds to Fig. 3.1 (6), then we obtain $\mathrm{III}^c, \mathrm{III}^d$ and IIIe.

Thus we have proved that every singular fiber is diffeomorphic to one of the fibers listed in the theorem.

In order to complete the proof of the first half of the theorem, we have only to show that if two singular fibers are diffeomorphic to each other, then they are C^∞ equivalent in the sense of Definition 1.1 (2), except for the two types of fibers I^0 and III^c.

Let $f_i : M_i \to N_i$, $i = 0, 1$, be proper C^∞ stable maps of orientable 4-manifolds into 3-manifolds. Let us take $y_i \in f_i(S(f_i)) \subset N_i$. Suppose that the singular fibers over y_0 and y_1 are diffeomorphic to each other.

If the singular fibers over y_0 and y_1 are of type I^0, then let $q_i \in S(f_i) \cap (f_i)^{-1}(y_i)$ be the unique singular point on the fibers. Since q_i are definite

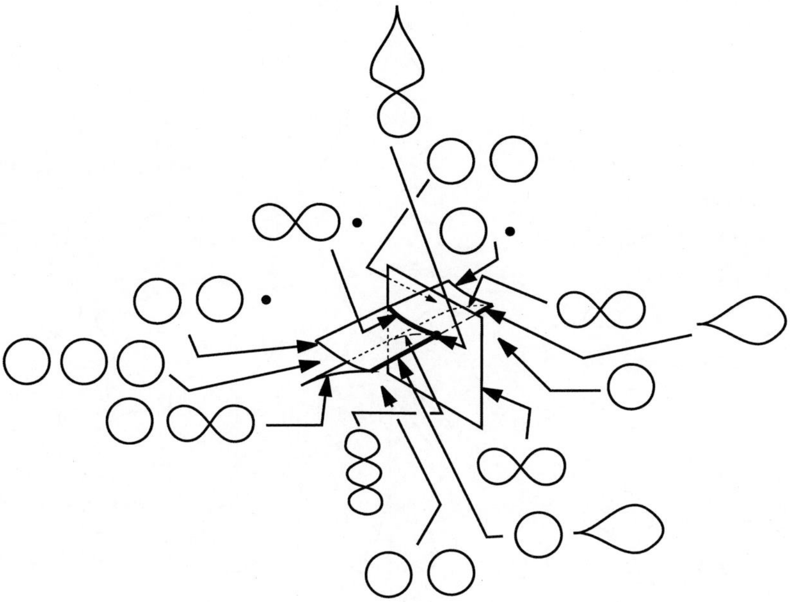

Fig. 3.7. Degeneration of fibers around the fiber of type III^b

fold points, there exist neighborhoods V_i of q_i in M_i, U_i of y_i in N_i and diffeomorphisms $\widetilde{\varphi}_0 : (V_0, q_0) \to (V_1, q_1)$ and $\varphi : (U_0, y_0) \to (U_1, y_1)$ which make the following diagram commutative:

$$\begin{array}{ccc} (V_0, q_0) & \xrightarrow{\widetilde{\varphi}_0} & (V_1, q_1) \\ {\scriptstyle f_0}\downarrow & & \downarrow{\scriptstyle f_1} \\ (U_0, y_0) & \xrightarrow{\varphi} & (U_1, y_1). \end{array}$$

Furthermore, by taking the neighborhoods sufficiently small, we may assume that $(U_i, U_i \cap f_i(S(f_i)))$ is as described in Fig. 3.1 (1), that V_i is a connected component of $(f_i)^{-1}(U_i)$, $U_i \cong \operatorname{Int} D^3$, $V_i \cong \operatorname{Int} D^4$, and $(f_i)^{-1}(y_i) \cap V_i = \{q_i\}$. Then the maps

$$f_i|_{(f_i)^{-1}(U_i) \smallsetminus V_i} : (f_i)^{-1}(U_i) \smallsetminus V_i \to U_i, \quad i = 0, 1$$

are proper submersions and their fibers are disjoint unions of the same number of copies of the circle. Hence, by Ehresmann's fibration theorem, the diffeomorphism $\widetilde{\varphi}_0 : (V_0, q_0) \to (V_1, q_1)$ extends to a diffeomorphism

$$\widetilde{\varphi} : ((f_0)^{-1}(U_0), (f_0)^{-1}(y_0)) \to ((f_1)^{-1}(U_1), (f_1)^{-1}(y_1))$$

so that the diagram (1.1) in Chap. 1 commutes. Hence, the fibers over y_0 and y_1 are equivalent.

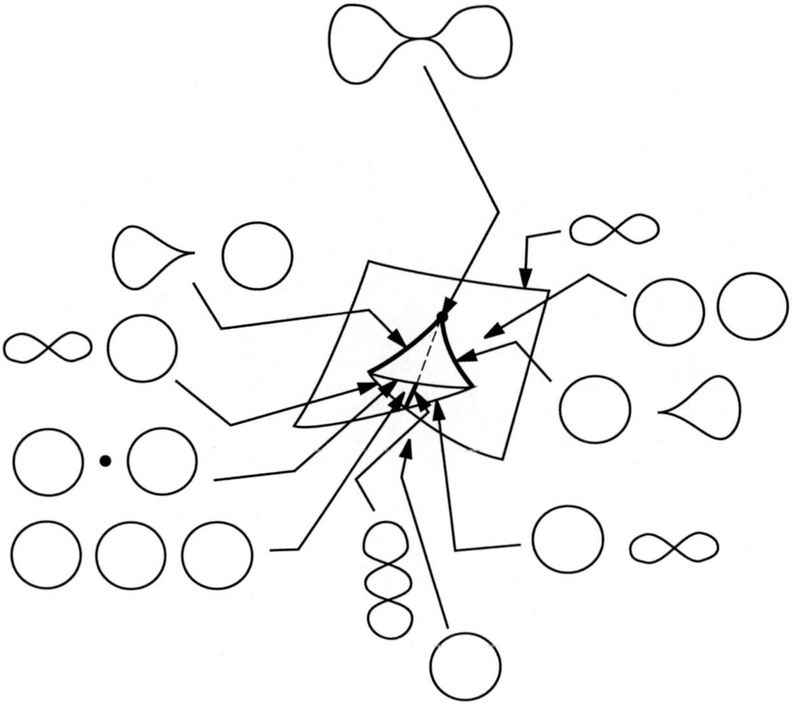

Fig. 3.8. Degeneration of fibers around the fiber of type IIIe

The same argument works when the fibers over y_0 and y_1 are of type III^c.

When the fibers over y_0 and y_1 are of type I^1, we can imitate the above argument for the case of I^0; however, we cannot take V_i to be a connected component of $(f_i)^{-1}(U_i)$, since the relevant singular points are indefinite fold points. So, we first take V_i sufficiently small, and then imitate the proof of Theorem 2.1. More precisely, we modify the diffeomorphisms $\widetilde{\varphi}_0 : V_0 \to V_1$ and $\varphi : U_0 \to U_1$, if necessary, by using self-diffeomorphisms of V_0 and U_0 corresponding to those defined by $(x, y, z, w) \mapsto (x, y, w, z)$ and $(X, Y, Z) \to (X, Y, -Z)$ respectively with respect to the coordinates as in Proposition 3.1 (i) so that the diffeomorphism

$$\widetilde{\varphi}_0 : (f_0)^{-1}(y_0) \cap V_0 \to (f_1)^{-1}(y_1) \cap V_1$$

extends to one between the whole fibers $(f_0)^{-1}(y_0)$ and $(f_1)^{-1}(y_1)$. Then we use the relative version of Ehresmann's fibration theorem to extend the diffeomorphism $\widetilde{\varphi}_0 : V_0 \to V_1$ to a fiber preserving diffeomorphism between $(f_0)^{-1}(U_0)$ and $(f_1)^{-1}(U_1)$. Hence, the fibers over y_0 and y_1 are equivalent.

The same argument works when the fiber over y_i contains exactly one singular point: namely, for the cases of II^a, III^d and IIIe.

Now suppose that the fibers over y_0 and y_1 are of type II^{00}. Then there exist neighborhoods U_i of y_i such that the sets $U_i \cap f_i(S(f_i))$ are as in Fig. 3.1 (2). In particular, there exists a diffeomorphism $\varphi : (U_0, y_0) \to (U_1, y_1)$ between the neighborhoods U_i of y_i such that

$$\varphi(U_0 \cap f_0(S(f_0))) = U_1 \cap f_1(S(f_1)).$$

Note that we can describe the degeneration of the fibers of f_i over U_i using Lemma 3.4 (for the case of II^3, see Fig. 3.5). Then we see that the diffeomorphism φ can be chosen so that it preserves the diffeomorphism types of the fibers: i.e., we may assume that $(f_0)^{-1}(y)$ is diffeomorphic to $(f_1)^{-1}(\varphi(y))$ for all $y \in U_0$. Put $(f_i)^{-1}(y_i) \cap S(f_i) = \{q_i, q_i'\}$, where q_i and q_i' are definite fold points. Then the multi-germs

$$\varphi \circ f_0 : ((f_0)^{-1}(U_0), \{q_0, q_0'\}) \to (U_1, y_1)$$

and

$$f_1 : ((f_1)^{-1}(U_1), \{q_1, q_1'\}) \to (U_1, y_1)$$

have the same discriminant set germ $(f_1(S(f_1)), y_1)$ and they satisfy the assumption of [11, (0.6) Theorem]. Hence they are right equivalent; i.e., there exists a diffeomorphism $\widetilde{\varphi}_0 : (V_0, \{q_0, q_0'\}) \to (V_1, \{q_1, q_1'\})$ between sufficiently small neighborhoods V_0 and V_1 of $\{q_0, q_0'\}$ and $\{q_1, q_1'\}$ respectively such that

$$f_1 \circ \widetilde{\varphi}_0 = \varphi \circ f_0 : (V_0, \{q_0, q_0'\}) \to (U_1, y_1)$$

(see also [61]). Then the rest of the proof is the same as that in the case of I^0.

When the fibers over y_0 and y_1 are of type II^{01}, put $(f_i)^{-1}(y_i) \cap S(f_i) = \{q_i, q_i'\}$, where q_i is a definite fold point and q_i' is an indefinite fold point. Then we can imitate the above argument to obtain a diffeomorphism φ between neighborhoods U_i of y_i and a diffeomorphism $\widetilde{\varphi}_0$ between neighborhoods V_i of $\{q_i, q_i'\}$ such that $f_1 \circ \widetilde{\varphi}_0 = \varphi \circ f_0$ on V_0. If we choose the diffeomorphism φ so that it preserves the diffeomorphism types of the fibers, then we see easily that the diffeomorphism $\widetilde{\varphi}_0$ between the local fibers $(f_0)^{-1}(y_0) \cap V_0$ and $(f_1)^{-1}(y_1) \cap V_1$ necessarily extends to one between the whole fibers $(f_0)^{-1}(y_0)$ and $(f_1)^{-1}(y_1)$; in other words, we do not need to modify $\widetilde{\varphi}_0$ or φ as in the proof of Theorem 2.1. Then the rest of the proof is the same as that in the case of I^1.

A similar argument works also in the cases of II^{11}, III^{000}, III^{001}, III^{011}, III^{111}, III^{0a} and III^{1a}.

When the fibers over y_0 and y_1 are of type II^2, we can use almost the same argument. The only difference is that we have to choose the diffeomorphism $\widetilde{\varphi}_0 : V_0 \to V_1$ so that the diffeomorphism

$$\widetilde{\varphi}_0 : (f_0)^{-1}(y_0) \cap V_0 \to (f_1)^{-1}(y_1) \cap V_1$$

between the local fibers extends to a diffeomorphism between the whole fibers $(f_0)^{-1}(y_0)$ and $(f_1)^{-1}(y_1)$. For this, we can use the self-diffeomorphisms of each of the neighborhoods of the indefinite fold points corresponding to those defined by $(x, y, z, w) \mapsto (x, y, \pm z, \pm w)$ with respect to the coordinates as in Proposition 3.1 (i). More precisely, we modify $\widetilde{\varphi}_0$ using these diffeomorphisms as we did in the case of I^1. Note that here, φ is chosen so that it preserves the diffeomorphism types of the fibers, and is fixed. Therefore, we cannot use the self-diffeomorphisms corresponding to those defined by $(x, y, z, w) \mapsto (x, y, \pm w, \pm z)$.

We can use similar arguments also in the cases of II^3, III^{02}, III^{03}, III^{12}, III^{13}, III^4, III^5, III^6, III^7, III^8 and III^b.

In the above argument, we note the following. When the fibers over y_0 and y_1 are of type III^{02}, III^{03}, III^{12}, III^{13}, III^4 or III^7, put $(f_i)^{-1}(y_i) \cap S(f_i) = \{q_i, q_i', q_i''\}$. We name them so that

$$\varphi(f_0(V_{0j} \cap S(f_0))) = f_1(V_{1j} \cap S(f_1))), \quad j = 1, 2, 3,$$

where V_i is the disjoint union of V_{i1}, V_{i2} and V_{i3} which are neighborhoods of q_i, q_i' and q_i'' respectively. Then we see easily that the correspondence $q_0 \mapsto q_1$, $q_0' \mapsto q_1'$, $q_0'' \mapsto q_1''$ coincides with that given by $\widetilde{\varphi}_0$ and extends to a diffeomorphism between the whole fibers $(f_0)^{-1}(y_0)$ and $(f_1)^{-1}(y_1)$, since φ preserves the diffeomorphism types of the fibers. (For the cases of II^2, II^3, III^5, III^6 and III^8, we do not need such an argument by virtue of their symmetries. For the case of III^h, we do not need it either because the two singular points contained in a fiber are of different types.) Therefore, we can apply the argument above.

The second half of the theorem is clear. This completes the proof of Theorem 3.5. □

Remark 3.7. Let $f : M \to N$ be a proper C^∞ stable map of an orientable 4-manifold M into a 3-manifold N and $\mathfrak{F}$ the type of one of the singular fibers appearing in Fig. 3.4. We define $\mathfrak{F}(f)$ to be the set of points $y \in N$ such that the fiber $f^{-1}(y)$ over y is equivalent to the disjoint union of $\mathfrak{F}$ and some copies of a fiber of the trivial circle bundle. As the above proof shows, each $\mathfrak{F}(f)$ is a submanifold of N, provided that it is nonempty, and its codimension is denoted by $\kappa(\mathfrak{F})$, which is called the *codimension* of the singular fiber of type $\mathfrak{F}$ (or the codimension of the disjoint union of $\mathfrak{F}$ and some copies of a fiber of the trivial circle bundle). See Fig. 3.4 for the codimension of each singular fiber. Note that the target manifold N is naturally stratified into these submanifolds.

Remark 3.8. As the proof of Theorem 3.5 shows, two singular fibers of proper C^∞ stable maps of orientable 4-manifolds into 3-manifolds are diffeomorphic if and only if they are C^∞ equivalent, except for the singular fibers of types I^0 and III^c.

Furthermore, we also have the following.

Corollary 3.9. *Two fibers of proper C^∞ stable maps of orientable 4-manifolds into 3-manifolds are C^∞ equivalent if and only if they are C^0 equivalent.*

Proof. We have only to prove the statement for arbitrary two fibers in the list given in Theorem 3.5. Suppose that two fibers are C^0 equivalent. Then the degenerations of the fibers around the singular fibers are also topologically equivalent, and their nearby fibers must be homeomorphic. It is not difficult to check that this implies that the two fibers are of the same C^∞ type. □

Remark 3.10. Recall that Damon [10] (see also [9]) has shown that for nice dimensions, two C^∞ stable map germs are topologically right-left equivalent if and only if they are smoothly right-left equivalent. The above corollary shows that this is also true for C^∞ stable map germs along fibers for the dimension pair $(4, 3)$, which is in the nice range, as long as the source manifold is orientable. (In fact, this is also true for the dimension pairs $(2, 1)$ and $(3, 2)$ without the orientability hypothesis. See Chap. 2 and Corollary 3.16 below.) Note that even for nice dimensions, this statement for map germs along fibers is not true in general. For example, we can construct two proper Morse functions of 8-dimensional manifolds such that one of them has the standard 7-dimensional sphere as its regular fibers, and that the other has a homotopy 7-sphere not diffeomorphic to the standard 7-sphere [33] as its regular fibers. Then the map germs along (nonsingular) fibers are topologically right-left equivalent, but not smoothly right-left equivalent.

Remark 3.11. Let us denote by $\mathbf{0}$ the smooth right-left equivalence class of a connected regular fiber. Furthermore, for a fiber of type $\mathfrak{F}$ and a positive integer n, we denote by $\mathfrak{F}_n$ the smooth right-left equivalence class of the fiber consisting of a fiber of type $\mathfrak{F}$ and some copies of a fiber of the trivial circle bundle such that the total number of connected components is equal to n. If we classify the fibers of proper C^∞ stable maps of orientable 4-manifolds into 3-manifolds up to homeomorphism in the sense of Definition 1.1 (2), then we get a smaller list than that given in Theorem 3.5. In fact, we have the following, where "$\approx$" means a homeomorphism:

(1) $\mathrm{I}^0_n \approx \mathrm{III}^c_n$ for $n \geq 1$,

(2) $\mathrm{I}^0_n \approx \mathrm{III}^c_n \approx \mathrm{III}^{0a}_n$ for $n \geq 2$,

(3) $\mathrm{I}^1_n \approx \mathrm{III}^b_n \approx \mathrm{III}^d_n \approx \mathrm{III}^e_n$ for $n \geq 1$,

(4) $\mathrm{I}^1_n \approx \mathrm{III}^b_n \approx \mathrm{III}^d_n \approx \mathrm{III}^e_n \approx \mathrm{III}^{1a}_n$ for $n \geq 2$,

(5) $\mathrm{III}^6_n \approx \mathrm{III}^8_n$ for $n \geq 1$,

(6) $\mathrm{II}^a_n \approx \mathbf{0}_n$ for $n \geq 1$.

Furthermore, it is not difficult to see that the above fibers exhaust all the repetitions of the homeomorphism types in the list of smooth right-left equivalence classes of fibers.

Remark 3.12. Suppose that a smooth map $f : M \to N$ between smooth manifolds is given. For two points $q, q' \in M$, we define $q \sim_f q'$ if $f(q) = f(q')$ and q and q' belong to the same connected component of an f-fiber. We define

$W_f = M/\sim_f$ to be the quotient space and $q_f : M \to W_f$ the quotient map. Then it is easy to see that there exists a unique continuous map $\overline{f} : W_f \to N$ such that the diagram

$$\begin{array}{ccc} M & \xrightarrow{f} & N \\ & {\scriptstyle q_f}\searrow \quad \nearrow{\scriptstyle \overline{f}} & \\ & W_f & \end{array}$$

is commutative. The space W_f or the above commutative diagram is called the *Stein factorization* of f (see [30]). It is known that if f is a topologically stable map, then W_f is a polyhedron and all the maps appearing in the above diagram are triangulable (for details, see [20]).

Kushner, Levine and Porto [28, 30] have determined the local structures of Stein factorizations of proper C^∞ stable maps of 3-manifolds into surfaces by using their classification of singular fibers. Similarly, by using our classification of singular fibers, we can determine the local structures of Stein factorizations of proper C^∞ stable maps of orientable 4-manifolds into 3-manifolds. For details, see [20].

Remark 3.13. In [63, 64], a similar classification of singular fibers of proper C^∞ stable maps of possibly nonorientable 4-manifolds into 3-manifolds is obtained.

3.2 Stable Maps of Surfaces and 3-Manifolds

In this section, let us mention similar classifications of singular fibers of proper C^∞ stable Morse functions on surfaces and those of proper C^∞ stable maps of 3-manifolds into surfaces. Let us begin by the following remark.

Remark 3.14. We can obtain a classification of singular fibers of proper C^∞ stable maps of orientable 3-manifolds into surfaces similar to Theorem 3.5. The list we get is nothing but the singular fibers with $\kappa = 1$ and 2 in Fig. 3.4. The list itself was already obtained by Kushner, Levine and Porto [28, 30], although they did not describe explicitly the equivalence relation for their classification.

In fact, we can easily get the following list of C^∞ right-left equivalence classes of singular fibers for proper C^∞ stable maps of (not necessarily orientable) 3-manifolds into surfaces. Details are left to the reader.

Theorem 3.15. *Let $f : M \to N$ be a proper C^∞ stable map of a 3-manifold M into a surface N. Then, every singular fiber of f is equivalent to the disjoint union of one of the fibers as in Fig.* 3.9 *and a finite number of copies of a fiber of the trivial circle bundle. Furthermore, no two fibers in the list are equivalent to each other even after taking the union with regular circle components.*

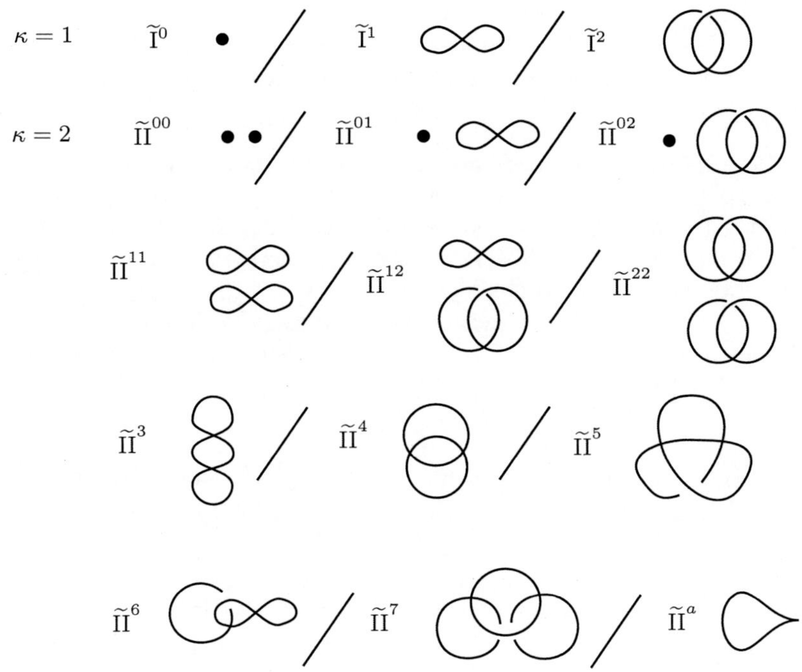

Fig. 3.9. List of singular fibers of proper C^∞ stable maps of 3-manifolds into surfaces

Note that the above list itself is mentioned in the introduction of [30]. As a corollary to Theorems 2.1 and 3.15, we get the following, which we can prove by an argument similar to that in the proof of Corollary 3.9. Details are left to the reader.

Corollary 3.16. *Let us consider two fibers of proper C^∞ stable Morse functions on surfaces, or two fibers of proper C^∞ stable maps of 3-manifolds into surfaces. Then, the following conditions are equivalent to each other.*

(1) *They are diffeomorphic.*
(2) *They are C^0 equivalent.*
(3) *They are C^∞ equivalent.*

We warn the reader that the fibers as depicted in Fig. 2.1 (2) and (3) (or the fibers $\widetilde{\mathrm{I}}^1$ and $\widetilde{\mathrm{I}}^2$) are homeomorphic to each other, although they are not C^0 equivalent nor diffeomorphic to each other. Compare these results with Remark 3.8 and Corollary 3.9.

As an important consequence of the above mentioned result, we show the following.

Corollary 3.17. *Let $f_0 : M_0 \to N_0$ and $f_1 : M_1 \to N_1$ be two proper C^∞ stable maps of surfaces into 1-dimensional manifolds. Then, the maps f_0 and f_1 are C^0 right-left equivalent if and only if they are C^∞ right-left equivalent.*

Proof. Suppose that f_0 and f_1 are C^0 right-left equivalent so that we have homeomorphisms $\widetilde{\varphi} : M_0 \to M_1$ and $\varphi : N_0 \to N_1$ satisfying $f_1 \circ \widetilde{\varphi} = \varphi \circ f_0$. Since $f_0(S(f_0))$ and $f_1(S(f_1))$ are discrete sets and φ sends $f_0(S(f_0))$ homeomorphically onto $f_1(S(f_1))$, we see that there exists a diffeomorphism $\psi : N_0 \to N_1$ which approximates φ such that $\psi|_{f_0(S(f_0))} = \varphi|_{f_0(S(f_0))}$.

Then by Corollary 3.16 together with the proof of Theorem 2.1, we see that for each point $y \in f_0(S(f_0))$, there exist a small neighborhood U_y of y in N_0 and a diffeomorphism $\widetilde{\psi}_y : (f_0)^{-1}(U_y) \to (f_1)^{-1}(U_{y'})$ such that the diagram

$$\begin{array}{ccc} ((f_0)^{-1}(U_y), (f_0)^{-1}(y)) & \xrightarrow{\widetilde{\psi}_y} & ((f_1)^{-1}(U_{y'}), (f_1)^{-1}(y')) \\ {\scriptstyle f_0}\downarrow & & \downarrow{\scriptstyle f_1} \\ (U_y, y) & \xrightarrow{\psi} & (U_{y'}, y') \end{array}$$

is commutative, where $y' = \psi(y)$ and $U_{y'} = \psi(U_y)$ is a neighborhood of y' in N_1. Here, we choose the diffeomorphism $\widetilde{\psi}_y$ so that it approximates $\widetilde{\varphi}|_{(f_0)^{-1}(U_y)}$.

Since the collection of homeomorphisms $\widetilde{\varphi}|_{(f_0)^{-1}(U_y)}$, $y \in f_0(S(f_0))$, extends to a homeomorphism $\widetilde{\varphi}$ such that $f_1 \circ \widetilde{\varphi} = \varphi \circ f_0$, the collection of diffeomorphisms $\widetilde{\psi}_y$, $y \in f_0(S(f_0))$, also extends to a homeomorphism $\widetilde{\psi}$ such that $f_1 \circ \widetilde{\psi} = \psi \circ f_0$.

Now it is well-known that two C^∞ S^1-bundles are C^0 equivalent if and only if they are C^∞ equivalent. This is true also for C^∞ bundles with fiber a union of finite copies of S^1. Hence the homeomorphism $\widetilde{\psi}$ above can be chosen to be a diffeomorphism. Hence, the C^∞ maps f_0 and f_1 are C^∞ right-left equivalent to each other. This completes the proof. □

The author does not know the answer to the following problem.

Problem 3.18. Let $f_0 : M_0 \to N_0$ and $f_1 : M_1 \to N_1$ be two proper C^∞ stable maps of orientable 4-manifolds into 3-manifolds (or two proper C^∞ stable maps of 3-manifolds into surfaces). If f_0 and f_1 are C^0 right-left equivalent, then are they C^∞ right-left equivalent?

For the above problem and Corollary 3.17, refer to [9, §4], for example. Note that there have been known a lot of examples of 4-manifold pairs which are mutually homeomorphic, but are not diffeomorphic. If the answer to the above problem is affirmative, then such 4-manifolds would not admit C^∞ stable maps that are C^0 right-left equivalent. (This suggests a possibility of constructing an invariant for 4-manifolds, from the viewpoint of singularity theory, that can detect the differentiable structures. For the construction of such an invariant, we have only to use the topological structures of stable maps.)

Let us recall the following example which is closely related to Problem 3.18.

Example 3.19. A smooth map $f : M \to N$ between manifolds of dimensions n and p with $n \geq p$ is called a *special generic map* if it has only definite fold points as its singularities (for example, see [43]). Sakuma and the author have found some examples of pairs (M_1, M_2) of smooth closed 4-dimensional manifolds with the following properties (see [44, 48, 49]).

(1) M_1 and M_2 are homeomorphic.
(2) M_1 admits a special generic map into $\mathbf{R}^3$.
(3) M_2 does not admit a special generic map into $\mathbf{R}^3$.

In other words, M_1 admits a stable map into $\mathbf{R}^3$ whose singular fibers are all of type I^0, II^{00} or III^{000}, while M_2 does not admit such a stable map, even though they are homeomorphic to each other.

4

Co-existence of Singular Fibers

Let $f : M \to N$ be a C^∞ stable map of a closed orientable 4-manifold into a 3-manifold. In this chapter, we consider a natural stratification of N induced by the equivalence classes of fibers of f, and obtain some relations among the numbers of singular fibers of codimension three.

Let $f : M \to N$ be a C^∞ stable map of a closed orientable 4-manifold M into a 3-manifold N and $\mathfrak{F}$ the equivalence class of one of the singular fibers appearing in Fig. 3.4. We define $\mathfrak{F}(f)$ to be the set of points $y \in N$ such that the fiber $f^{-1}(y)$ over y is equivalent to the union of $\mathfrak{F}$ and some copies of a fiber of the trivial circle bundle. Furthermore, we define $\mathfrak{F}_{\mathrm{o}}(f)$ (resp. $\mathfrak{F}_{\mathrm{e}}(f)$) to be the subset of $\mathfrak{F}(f)$ consisting of the points $y \in N$ such that $b_0(f^{-1}(y))$ is odd (resp. even), where b_0 denotes the number of connected components. We denote the closures of $\mathfrak{F}(f)$, $\mathfrak{F}_{\mathrm{o}}(f)$, and $\mathfrak{F}_{\mathrm{e}}(f)$ in N by $\overline{\mathfrak{F}(f)}$, $\overline{\mathfrak{F}_{\mathrm{o}}(f)}$, and $\overline{\mathfrak{F}_{\mathrm{e}}(f)}$, respectively. It is easy to see that each of $\overline{\mathfrak{F}(f)}$, $\overline{\mathfrak{F}_{\mathrm{o}}(f)}$, or $\overline{\mathfrak{F}_{\mathrm{e}}(f)}$ is a $(3-\kappa)$-dimensional subcomplex of N, where κ is the codimension of $\mathfrak{F}$. In particular, if the codimension κ is equal to two, then $\overline{\mathfrak{F}_{\mathrm{o}}(f)}$ and $\overline{\mathfrak{F}_{\mathrm{e}}(f)}$ are finite graphs embedded in N. Their vertices correspond to points over which lies a singular fiber with $\kappa = 3$. For a singular fiber $\mathfrak{F}'$ of $\kappa = 3$, the degree of the vertex corresponding to $\mathfrak{F}'_{\mathrm{o}}(f)$ (or $\mathfrak{F}'_{\mathrm{e}}(f)$) in the graph $\overline{\mathfrak{F}_{\mathrm{o}}(f)}$ is given in Table 4.1, which can be obtained by using the description of nearby fibers as in Fig. 3.6–3.8. Note that the degrees in the graph $\overline{\mathfrak{F}_{\mathrm{e}}(f)}$ can be obtained by interchanging $\mathfrak{F}'_{\mathrm{o}}(f)$ with $\mathfrak{F}'_{\mathrm{e}}(f)$ in the table.

In the following, for a finite set X, we denote by $|X|$ the number of its elements. Since the sum of the degrees over all vertices is always an even number for any finite graph,[1] we obtain the following.

Proposition 4.1. *Let $f : M \to N$ be a C^∞ stable map of a closed orientable 4-manifold into a 3-manifold. Then the following numbers are always even.*

(1) $|\mathrm{III}^{000}(f)| + |\mathrm{III}^{001}(f)| + |\mathrm{III}^{0a}_{\mathrm{e}}(f)| + |\mathrm{III}^{c}_{\mathrm{e}}(f)|$.

(2) $|\mathrm{III}^{000}(f)| + |\mathrm{III}^{001}(f)| + |\mathrm{III}^{0a}_{\mathrm{o}}(f)| + |\mathrm{III}^{c}_{\mathrm{o}}(f)|$.

[1] This is due to Euler and is said to be the oldest theorem in the graph theory.

Table 4.1. Degree of each vertex in the graphs

	$\overline{\mathrm{II}^{00}_{\mathrm{o}}(f)}$	$\overline{\mathrm{II}^{01}_{\mathrm{o}}(f)}$	$\overline{\mathrm{II}^{11}_{\mathrm{o}}(f)}$	$\overline{\mathrm{II}^{2}_{\mathrm{o}}(f)}$	$\overline{\mathrm{II}^{3}_{\mathrm{o}}(f)}$	$\overline{\mathrm{II}^{a}_{\mathrm{o}}(f)}$
$\mathrm{III}^{000}_{\mathrm{o}}(f)$	3	0	0	0	0	0
$\mathrm{III}^{000}_{\mathrm{e}}(f)$	3	0	0	0	0	0
$\mathrm{III}^{001}_{\mathrm{o}}(f)$	1	2	0	0	0	0
$\mathrm{III}^{001}_{\mathrm{e}}(f)$	1	2	0	0	0	0
$\mathrm{III}^{011}_{\mathrm{o}}(f)$	0	2	1	0	0	0
$\mathrm{III}^{011}_{\mathrm{e}}(f)$	0	2	1	0	0	0
$\mathrm{III}^{111}_{\mathrm{o}}(f)$	0	0	3	0	0	0
$\mathrm{III}^{111}_{\mathrm{e}}(f)$	0	0	3	0	0	0
$\mathrm{III}^{02}_{\mathrm{o}}(f)$	0	2	0	1	0	0
$\mathrm{III}^{02}_{\mathrm{e}}(f)$	0	2	0	1	0	0
$\mathrm{III}^{03}_{\mathrm{o}}(f)$	0	4	0	0	1	0
$\mathrm{III}^{03}_{\mathrm{e}}(f)$	0	0	0	0	1	0
$\mathrm{III}^{12}_{\mathrm{o}}(f)$	0	0	2	1	0	0
$\mathrm{III}^{12}_{\mathrm{e}}(f)$	0	0	2	1	0	0
$\mathrm{III}^{13}_{\mathrm{o}}(f)$	0	0	4	0	1	0
$\mathrm{III}^{13}_{\mathrm{e}}(f)$	0	0	0	0	1	0
$\mathrm{III}^{4}_{\mathrm{o}}(f)$	0	0	0	3	0	0
$\mathrm{III}^{4}_{\mathrm{e}}(f)$	0	0	1	2	0	0
$\mathrm{III}^{5}_{\mathrm{o}}(f)$	0	0	0	3	0	0
$\mathrm{III}^{5}_{\mathrm{e}}(f)$	0	0	0	3	0	0
$\mathrm{III}^{6}_{\mathrm{o}}(f)$	0	0	0	3	3	0
$\mathrm{III}^{6}_{\mathrm{e}}(f)$	0	0	0	0	0	0
$\mathrm{III}^{7}_{\mathrm{o}}(f)$	0	0	0	4	1	0
$\mathrm{III}^{7}_{\mathrm{e}}(f)$	0	0	0	0	1	0
$\mathrm{III}^{8}_{\mathrm{o}}(f)$	0	0	0	0	6	0
$\mathrm{III}^{8}_{\mathrm{e}}(f)$	0	0	0	0	0	0
$\mathrm{III}^{0a}_{\mathrm{o}}(f)$	0	1	0	0	0	1
$\mathrm{III}^{0a}_{\mathrm{e}}(f)$	1	0	0	0	0	1
$\mathrm{III}^{1a}_{\mathrm{o}}(f)$	0	0	1	0	0	1
$\mathrm{III}^{1a}_{\mathrm{e}}(f)$	0	1	0	0	0	1
$\mathrm{III}^{b}_{\mathrm{o}}(f)$	0	0	0	1	0	1
$\mathrm{III}^{b}_{\mathrm{e}}(f)$	0	1	0	0	0	1
$\mathrm{III}^{c}_{\mathrm{o}}(f)$	0	0	0	0	0	2
$\mathrm{III}^{c}_{\mathrm{e}}(f)$	1	0	0	0	0	0
$\mathrm{III}^{d}_{\mathrm{o}}(f)$	0	0	0	0	1	2
$\mathrm{III}^{d}_{\mathrm{e}}(f)$	0	0	0	0	0	0
$\mathrm{III}^{e}_{\mathrm{o}}(f)$	0	0	0	1	0	0
$\mathrm{III}^{e}_{\mathrm{e}}(f)$	0	0	0	0	0	2

(3) $|\mathrm{III}^{0a}_{\mathrm{o}}(f)| + |\mathrm{III}^{1a}_{\mathrm{e}}(f)| + |\mathrm{III}^{b}_{\mathrm{e}}(f)|$.

(4) $|\mathrm{III}^{0a}_{\mathrm{e}}(f)| + |\mathrm{III}^{1a}_{\mathrm{o}}(f)| + |\mathrm{III}^{b}_{\mathrm{o}}(f)|$.

(5) $|\mathrm{III}^{011}(f)| + |\mathrm{III}^{111}(f)| + |\mathrm{III}^{4}_{\mathrm{e}}(f)| + |\mathrm{III}^{1a}_{\mathrm{o}}(f)|$.

(6) $|\mathrm{III}^{011}(f)| + |\mathrm{III}^{111}(f)| + |\mathrm{III}^{4}_{\mathrm{o}}(f)| + |\mathrm{III}^{1a}_{\mathrm{e}}(f)|$.

(7) $|\mathrm{III}^{02}(f)| + |\mathrm{III}^{12}(f)| + |\mathrm{III}^{4}_{\mathrm{o}}(f)| + |\mathrm{III}^{5}(f)| + |\mathrm{III}^{6}_{\mathrm{o}}(f)| + |\mathrm{III}^{b}_{\mathrm{o}}(f)| + |\mathrm{III}^{e}_{\mathrm{o}}(f)|$.

(8) $|\mathrm{III}^{02}(f)| + |\mathrm{III}^{12}(f)| + |\mathrm{III}^{4}_{\mathrm{e}}(f)| + |\mathrm{III}^{5}(f)| + |\mathrm{III}^{6}_{\mathrm{e}}(f)| + |\mathrm{III}^{b}_{\mathrm{e}}(f)| + |\mathrm{III}^{e}_{\mathrm{e}}(f)|$.

(9) $|\mathrm{III}^{03}(f)| + |\mathrm{III}^{13}(f)| + |\mathrm{III}^{6}_{\mathrm{o}}(f)| + |\mathrm{III}^{7}(f)| + |\mathrm{III}^{d}_{\mathrm{o}}(f)|$.

(10) $|\mathrm{III}^{03}(f)| + |\mathrm{III}^{13}(f)| + |\mathrm{III}^{6}_{\mathrm{e}}(f)| + |\mathrm{III}^{7}(f)| + |\mathrm{III}^{d}_{\mathrm{e}}(f)|$.

(11) $|\mathrm{III}^{0a}(f)| + |\mathrm{III}^{1a}(f)| + |\mathrm{III}^{b}(f)|$.

In fact, items (1)–(10) of the above proposition correspond to the graphs $\overline{\mathrm{II}^{00}_{\mathrm{o}}(f)}$, $\overline{\mathrm{II}^{00}_{\mathrm{e}}(f)}$, $\overline{\mathrm{II}^{01}_{\mathrm{o}}(f)}$, $\overline{\mathrm{II}^{01}_{\mathrm{e}}(f)}$, $\overline{\mathrm{II}^{11}_{\mathrm{o}}(f)}$, $\overline{\mathrm{II}^{11}_{\mathrm{e}}(f)}$, $\overline{\mathrm{II}^{2}_{\mathrm{o}}(f)}$, $\overline{\mathrm{II}^{2}_{\mathrm{e}}(f)}$, $\overline{\mathrm{II}^{3}_{\mathrm{o}}(f)}$, and $\overline{\mathrm{II}^{3}_{\mathrm{e}}(f)}$ respectively. Item (11) corresponds to both $\overline{\mathrm{II}^{a}_{\mathrm{o}}(f)}$ and $\overline{\mathrm{II}^{a}_{\mathrm{e}}(f)}$.

Eliminating the terms of the forms $|\mathfrak{F}_{\mathrm{o}}(f)|$ and $|\mathfrak{F}_{\mathrm{e}}(f)|$, we obtain the following.

Corollary 4.2. *Let $f : M \to N$ be a C^∞ stable map of a closed orientable 4-manifold into a 3-manifold. Then the following numbers are always even.*

(1) $|\mathrm{III}^{0a}(f)| + |\mathrm{III}^{c}(f)|$.

(2) $|\mathrm{III}^{0a}(f)| + |\mathrm{III}^{1a}(f)| + |\mathrm{III}^{b}(f)|$.

(3) $|\mathrm{III}^{4}(f)| + |\mathrm{III}^{1a}(f)|$.

(4) $|\mathrm{III}^{4}(f)| + |\mathrm{III}^{6}(f)| + |\mathrm{III}^{b}(f)| + |\mathrm{III}^{e}(f)|$.

(5) $|\mathrm{III}^{6}(f)| + |\mathrm{III}^{d}(f)|$.

Remark 4.3. It is easy to see that the five numbers appearing in Corollary 4.2 are all even if and only if the following five hold.

(1) $|\mathrm{III}^{0a}(f)| \equiv |\mathrm{III}^{c}(f)| \pmod 2$.

(2) $|\mathrm{III}^{1a}(f)| \equiv |\mathrm{III}^{4}(f)| \pmod 2$.

(3) $|\mathrm{III}^{6}(f)| \equiv |\mathrm{III}^{d}(f)| \pmod 2$.

(4) $|\mathrm{III}^{b}(f)| \equiv |\mathrm{III}^{4}(f)| + |\mathrm{III}^{c}(f)| \pmod 2$.

(5) $|\mathrm{III}^{c}(f)| + |\mathrm{III}^{d}(f)| + |\mathrm{III}^{e}(f)| \equiv 0 \pmod 2$.

Note that the left hand side of congruence (5) is nothing but the total number of swallowtails. Note also that item (11) of Proposition 4.1 represents the number of cuspidal intersections as in Fig. 3.1 (5).

Remark 4.4. Adding items (2), (3), (6), (8) and (10) of Proposition 4.1, we obtain

$$\begin{aligned}&|\mathrm{III}^{000}(f)|+|\mathrm{III}^{001}(f)|+|\mathrm{III}^{011}(f)|+|\mathrm{III}^{111}(f)|+|\mathrm{III}^{02}(f)|+|\mathrm{III}^{03}(f)|\\&\quad+|\mathrm{III}^{12}(f)|+|\mathrm{III}^{13}(f)|+|\mathrm{III}^{4}(f)|+|\mathrm{III}^{5}(f)|+|\mathrm{III}^{7}(f)|\\&\quad+|\mathrm{III}^{c}_{\mathrm{o}}(f)|+|\mathrm{III}^{d}_{\mathrm{e}}(f)|+|\mathrm{III}^{e}_{\mathrm{e}}(f)|\equiv 0 \pmod 2.\end{aligned}$$

This and congruence (1) of Remark 4.3 have also been obtained in [20] by using methods different from ours.

Remark 4.5. By using the same method, we can obtain similar co-existence results for singular fibers of proper C^∞ stable maps of closed 3-manifolds into surfaces. More precisely, using the notation introduced in Theorem 3.15, we have the following.

(1) $|\widetilde{\mathrm{II}}^{01}(f)|+|\widetilde{\mathrm{II}}^{a}_{\mathrm{e}}(f)|\equiv 0 \pmod 2$.

(2) $|\widetilde{\mathrm{II}}^{01}(f)|+|\widetilde{\mathrm{II}}^{a}_{\mathrm{o}}(f)|\equiv 0 \pmod 2$.

(3) $|\widetilde{\mathrm{II}}^{02}(f)|+|\widetilde{\mathrm{II}}^{12}(f)|+|\widetilde{\mathrm{II}}^{6}(f)|\equiv 0 \pmod 2$.

Details are left to the reader (compare this with Table 9.2 of Chap. 9).

We end this chapter by posing a problem.

Problem 4.6. Let $\mathcal{S}$ be the $\mathbf{Z}_2$ vector space consisting of 38-tuples of elements of $\mathbf{Z}_2$ such that the congruences in Proposition 4.1 hold, where each of the 38 components corresponds to $|\mathrm{III}^{000}_{\mathrm{o}}(f)|$, $|\mathrm{III}^{000}_{\mathrm{e}}(f)|$, etc. Then, for an arbitrary element of $\mathcal{S}$, does there exist a C^∞ stable map of some closed orientable 4-manifold into some 3-manifold which realizes it as the parities of the numbers of corresponding singular fibers? In other words, do the congruences in Proposition 4.1 exhaust all the possible relations among the parities of the numbers of singular fibers of the form $\mathfrak{F}_{\mathrm{o}}(f)$ or $\mathfrak{F}_{\mathrm{e}}(f)$?

5

Euler Characteristic of the Source 4-Manifold

In this chapter, using the co-existence results for singular fibers obtained in the previous chapter, we study the relationship between the number of singular fibers of a certain type and the Euler characteristic of the source 4-manifold. In the following, χ will denote the Euler characteristic.

Let $f : M \to N$ be a C^∞ stable map of a closed orientable 4-manifold into a 3-manifold. Set

$$\begin{aligned}\mathbf{O}_\mathrm{o}(f) &= \{y \in N \smallsetminus f(S(f)) \,:\, b_0(f^{-1}(y)) \equiv 1 \pmod 2\},\\ \mathbf{O}_\mathrm{e}(f) &= \{y \in N \smallsetminus f(S(f)) \,:\, b_0(f^{-1}(y)) \equiv 0 \pmod 2\}.\end{aligned}$$

It is easy to see that they are disjoint open sets of N. Furthermore, since M is compact, the closure $\overline{\mathbf{O}_\mathrm{o}(f)}$ of $\mathbf{O}_\mathrm{o}(f)$ is compact. Let y and y' be points in N belonging to adjacent regions of $N \smallsetminus f(S(f))$. Since M is orientable, the difference between the numbers of components of the fibers over y and y' is always equal to one. Hence, we have

$$\overline{\mathbf{O}_\mathrm{o}(f)} \cap \overline{\mathbf{O}_\mathrm{e}(f)} = \partial\mathbf{O}_\mathrm{o}(f) = \partial\mathbf{O}_\mathrm{e}(f) = f(S(f)),$$

where for a subset X of a topological space, ∂X denotes $\overline{X} \smallsetminus \operatorname{Int} X$. In other words, $(N, f(S(f)))$ is two colorable in the sense of [36] (see also [35]).

Note that the map $f|_{S(f)} : S(f) \to N$ is a topologically stable singular surface in the sense of [36]. Then, for each cross cap $y \in f(S(f))$, which corresponds to a swallowtail point of f, we can define the index $\operatorname{Ind}_f(y) \in \{0, 1\}$ by using the coloring $(\mathbf{O}_\mathrm{o}(f), \mathbf{O}_\mathrm{e}(f))$ of $(N, f(S(f)))$. More precisely, it is defined as in Fig. 5.1 (for details, see [36]).

Then by Szűcs' formula [55] (see also [36, 37]), we have

$$T(f(S(f))) + \sum_y \operatorname{Ind}_f(y) \equiv \chi(S(f)) \pmod 2, \tag{5.1}$$

where y runs through the cross caps of $f(S(f))$ corresponding to Fig. 3.1 (6), and $T(f(S(f)))$ denotes the number of triple points of $f(S(f))$ corresponding

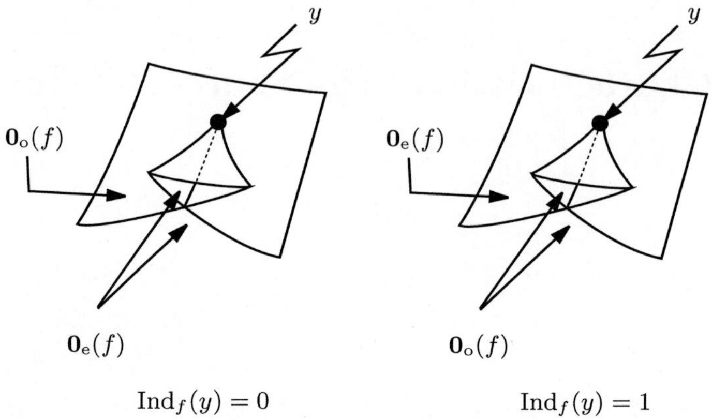

Fig. 5.1. Index of a cross cap

to Fig. 3.1 (4). On the other hand, by using the degenerations of the fibers around the singular fibers corresponding to swallowtails as in Fig. 3.8, we obtain the following:

$$\mathrm{Ind}_f(y) = \begin{cases} 0, & \text{if } y \in \mathrm{III}^c_{\rm o}(f) \cup \mathrm{III}^d_{\rm o}(f) \cup \mathrm{III}^c_{\rm e}(f), \\ 1, & \text{if } y \in \mathrm{III}^c_{\rm e}(f) \cup \mathrm{III}^d_{\rm e}(f) \cup \mathrm{III}^e_{\rm o}(f). \end{cases}$$

Hence, applying (5.1), we have

$$\begin{aligned} &|\mathrm{III}^{000}(f)| + |\mathrm{III}^{001}(f)| + |\mathrm{III}^{011}(f)| + |\mathrm{III}^{111}(f)| + |\mathrm{III}^{02}(f)| + |\mathrm{III}^{03}(f)| \\ &\quad + |\mathrm{III}^{12}(f)| + |\mathrm{III}^{13}(f)| + |\mathrm{III}^{4}(f)| + |\mathrm{III}^{5}(f)| + |\mathrm{III}^{6}(f)| + |\mathrm{III}^{7}(f)| \\ &\quad + |\mathrm{III}^{8}(f)| + |\mathrm{III}^c_{\rm e}(f)| + |\mathrm{III}^d_{\rm e}(f)| + |\mathrm{III}^e_{\rm o}(f)| \equiv \chi(S(f)) \pmod 2. \end{aligned}$$

On the other hand, adding items (1), (3), (5), (7), (9) and (11) in Proposition 4.1, we obtain

$$\begin{aligned} &|\mathrm{III}^{000}(f)| + |\mathrm{III}^{001}(f)| + |\mathrm{III}^{011}(f)| + |\mathrm{III}^{111}(f)| + |\mathrm{III}^{02}(f)| + |\mathrm{III}^{03}(f)| \\ &\quad + |\mathrm{III}^{12}(f)| + |\mathrm{III}^{13}(f)| + |\mathrm{III}^{4}(f)| + |\mathrm{III}^{5}(f)| + |\mathrm{III}^{7}(f)| + |\mathrm{III}^c_{\rm e}(f)| \\ &\quad + |\mathrm{III}^d_{\rm o}(f)| + |\mathrm{III}^e_{\rm o}(f)| \equiv 0 \pmod 2. \end{aligned}$$

Adding the above two congruences, we obtain

$$|\mathrm{III}^{6}(f)| + |\mathrm{III}^{8}(f)| + |\mathrm{III}^{d}(f)| \equiv \chi(S(f)) \pmod 2.$$

Since $|\mathrm{III}^{6}(f)| \equiv |\mathrm{III}^{d}(f)| \pmod 2$ by Corollary 4.2 (5), we get

$$|\mathrm{III}^{8}(f)| \equiv \chi(S(f)) \pmod 2.$$

Since we always have

$$\chi(S(f)) \equiv \chi(M) \pmod 2$$

by [14, 45], we finally obtain the following theorem, which can be regarded as a 4-dimensional version of Corollary 2.4.

Theorem 5.1. *Let $f : M \to N$ be a C^∞ stable map of a closed orientable 4-manifold into a 3-manifold. Then we have*

$$\chi(M) \equiv |\mathrm{III}^8(f)| \pmod 2.$$

Remark 5.2. The above theorem holds also for C^∞ stable maps of closed (not necessarily orientable) 4-manifolds into 3-manifolds such that every fiber has an orientable neighborhood. For example, a smooth map $f : M \to N$ between manifolds satisfies this property if $f^*w_1(N) = w_1(M)$, where w_1 denotes the first Stiefel-Whitney class. Such a map f is said to be *orientable* in [4].

Remark 5.3. The results in the previous and the present chapters can be generalized to C^∞ stable maps of possibly nonorientable closed 4-manifolds into 3-manifolds. For details, see [63, 64] (see also Remark 3.13).

Remark 5.4. A result corresponding to Remark 2.7 does not hold for singular fibers of types III^* for C^∞ stable maps of 4-manifolds into 3-manifolds. This is because we can increase the number of fibers of a given type of codimension three as much as we want. For details, see Remark 3.6.

6

Examples of Stable Maps of 4-Manifolds

In this chapter, we give explicit examples of C^∞ stable maps of 4-manifolds into $\mathbf{R}^3$. Note that there have already been known some explicit examples of such stable maps that have only definite fold points as their singularities, i.e. special generic maps (see [51, 43, 44, 48, 49]). Such maps have singular fibers of types I^0, II^{00}, and III^{000}, and have no other singular fibers. Furthermore, the source 4-manifolds of such maps always have even Euler characteristics. Here we construct more complicated maps having a singular fiber of type III^8 such that the source 4-manifold has odd Euler characteristic.

Since $(4, 3)$ is a nice dimension pair, given a 4-manifold M and a 3-manifold N, we know that there are plenty of C^∞ stable maps of M into N. However, there has been known no systematic method to *construct an explicit example* of such a map. In this chapter we will introduce a (rather straightforward) method to construct such maps by pasting elementary parts.

Let us first construct a C^∞ stable map $f : \mathbf{C}P^2 \sharp 2\overline{\mathbf{C}P^2} \to \mathbf{R}^3$ which satisfies the following properties.

(1) The map f has only fold points as its singularities.
(2) The singular set $S(f)$ is the union of three 2-sphere components consisting of definite fold points and a projective plane component consisting of indefinite fold points.
(3) The discriminant set $f(S(f))$ is a disjoint union of three embedded 2-spheres and the Boy surface in $\mathbf{R}^3$ (see Fig. 6.1).
(4) The fibers of f can be completely described (details will be given in Fig. 6.2).

Recall that the Boy surface P, which is the image of an immersion $\mathbf{R}P^2 \looparrowright \mathbf{R}^3$, is constructed by attaching a 2-disk as in the right hand side of Fig. 6.1 to the image of an immersion of the Möbius band as in the left hand side of Fig. 6.1, from the front side.

Note that $\mathbf{R}^3 \setminus P$ consists exactly of two regions. Let S_0 be a 2-sphere embedded in the unbounded region of $\mathbf{R}^3 \setminus P$ such that the bounded region of $\mathbf{R}^3 \setminus S_0$ contains P. Furthermore, let S_1 and S_2 be two disjoint concentric

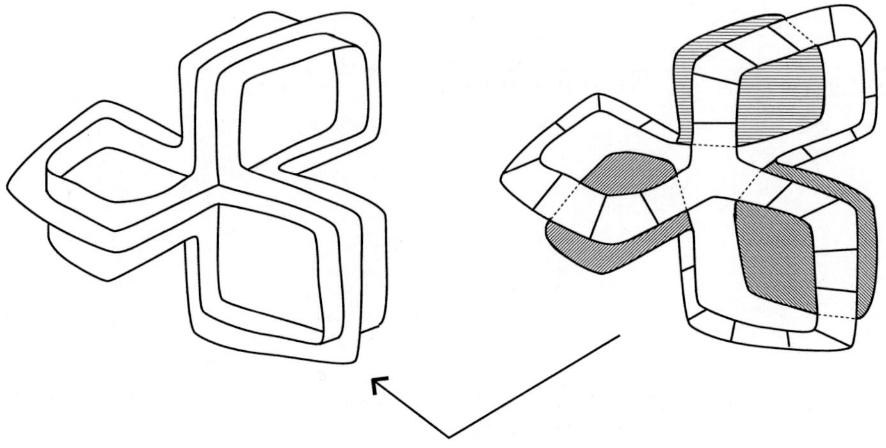

Fig. 6.1. Boy surface

2-spheres embedded in the bounded region of $\mathbf{R}^3 \setminus P$ such that S_2 is contained in the bounded region of $\mathbf{R}^3 \setminus S_1$. Note that $\mathbf{R}^3 \setminus (P \cup S_0 \cup S_1 \cup S_2)$ consists exactly of five regions and that $P \cup S_0 \cup S_1 \cup S_2$ naturally induces a stratification of $\mathbf{R}^3$: we have five strata of dimension three, seven strata of dimension two, three strata of dimension one, and one stratum of dimension zero. Let us denote by A^i_j the strata of dimension i. We enumerate them as follows (see Fig. 6.2):

(1) the closure of A^2_j contains $A^0_1 \cup A^1_j$, $j = 1, 2, 3$, and the closure of A^2_4 contains $A^0_1 \cup A^1_1 \cup A^1_2 \cup A^1_3$,

(2) $A^2_5 = S_0$, $A^2_6 = S_1$, $A^2_7 = S_2$,

(3) A^3_1 is the unbounded region of $\mathbf{R}^3 \setminus S_0$,

(4) A^3_2 is the region between S_0 and the Boy surface,

(5) A^3_3 is the region between the Boy surface and S_1,

(6) A^3_4 is the region between S_1 and S_2, and

(7) A^3_5 is the bounded region of $\mathbf{R}^3 \setminus S_2$.

We shall construct a fold map $f : \mathbf{C}P^2 \sharp 2\overline{\mathbf{C}P^2} \to \mathbf{R}^3$ such that $f(S_0(f)) = S_0 \cup S_1 \cup S_2$ and $f(S_1(f)) = P$, where a *fold map* is a smooth map with only fold points as its singularities. In particular, $S_0(f)$ is diffeomorphic to the disjoint union of three 2-spheres and $S_1(f)$ is diffeomorphic to $\mathbf{R}P^2$.

Over the points on each stratum we put fibers as depicted in Fig. 6.2, where the lower figure depicts a part of the 2-disk (contained in P) as in the right hand side of Fig. 6.1 together with parts of S_1 and S_2, which sit inside the bounded region of $\mathbf{R}^3 \setminus P$. It is easy to see that the regular parts of the fibers

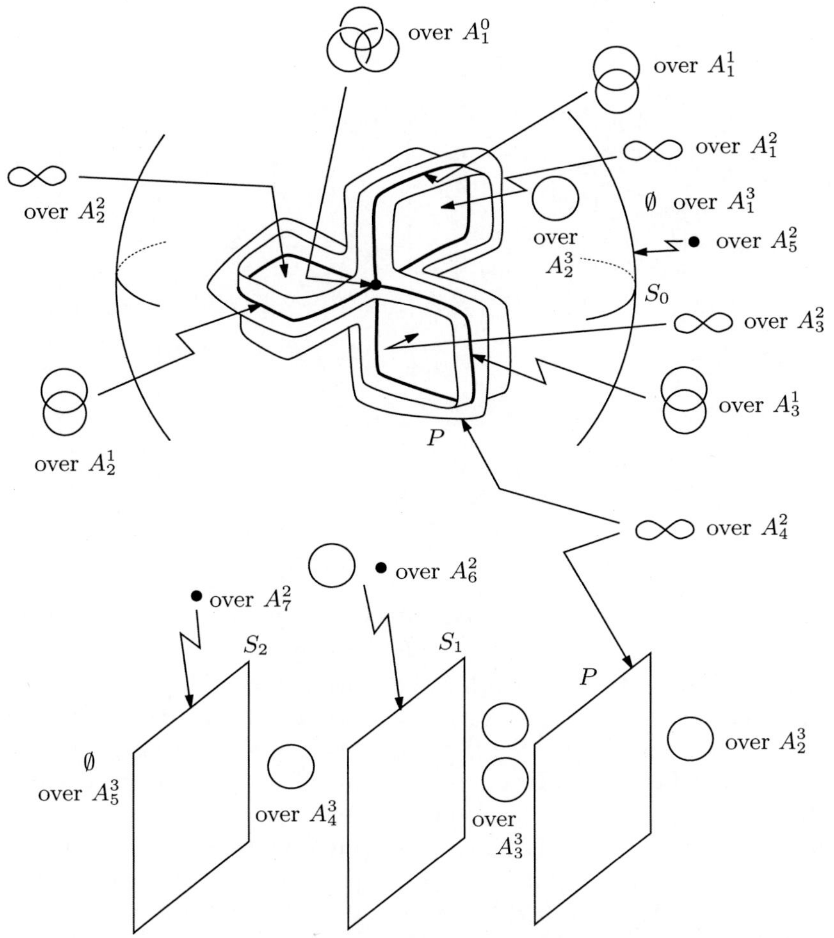

Fig. 6.2. Fibers over the points in $\mathbf{R}^3$

can be oriented consistently. Hence, if such a smooth map is constructed, then the source 4-manifold will be orientable.

Let $N(A_1^0)$ be a small closed disk neighborhood of the zero dimensional stratum A_1^0 such that its boundary two sphere is transverse to the other strata. Let $N(A_j^1) \cong D^2 \times [0,1]$ denote the closure of $\widetilde{N}(A_j^1) \setminus N(A_1^0)$, where $\widetilde{N}(A_j^1)$ is a small tubular neighborhood of the 1-dimensional stratum A_j^1 such that its boundary is transverse to the strata of higher dimensions ($j = 1, 2, 3$). We may assume that $N(A_j^1) \cong D^2 \times [0,1]$ is attached to $N(A_1^0)$ along $D^2 \times \{0,1\}$ and that $N(A_1^0) \cup N(A_1^1) \cup N(A_2^1) \cup N(A_3^1)$ is a regular neighborhood of $A_1^0 \cup A_1^1 \cup A_2^1 \cup A_3^1$ in $\mathbf{R}^3$. Similarly, we construct $N(A_j^2)$, $j = 1, 2, \ldots, 7$, and $N(A_j^3)$, $j = 1, 2, \ldots, 5$, so that the family of closed sets $\{N(A_j^i)\}_{0 \leq i \leq 3}$

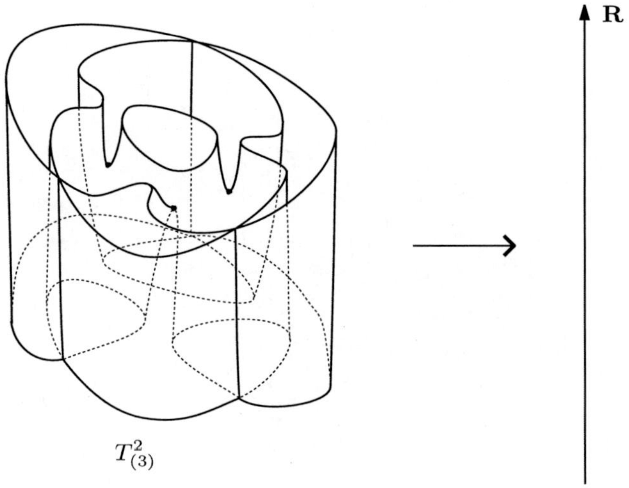

Fig. 6.3. A 2-parameter deformation of Morse functions on $T^2_{(3)}$

covers $\mathbf{R}^3$ and that distinct members intersect only along their boundaries. Furthermore, we put $\widehat{A}^i_j = A^i_j \cap N(A^i_j)$. We may assume that the natural projection $N(A^i_j) \to \widehat{A}^i_j$ is a smooth $(3-i)$-disk bundle.

Let us now construct a closed orientable 4-manifold M and a C^∞ stable map $f : M \to \mathbf{R}^3$ such that $f(S(f))$ and the fibers are as depicted in Fig. 6.2. Our strategy is to first construct compact 4-manifolds M^i_j and smooth maps $f^i_j : M^i_j \to N(A^i_j)$, and then glue them together.

As we have noted in Remark 3.6, we can construct a compact orientable 4-manifold M^0_1 and a smooth map $f^0_1 : M^0_1 \to \mathbf{R}^3$ which has only fold points as its singularities such that $f^0_1(M^0_1) = N(A^0_1)$ and that the fibers are consistent with Fig. 6.2 (see also Fig. 3.6). In our case, M^0_1 is diffeomorphic to $T^2_{(3)} \times D^2$, where for a surface F, we denote by $F_{(\ell)}$ the surface obtained from F by taking off ℓ open disks whose closures do not intersect each other, and T^2 denotes the 2-dimensional torus. Such a map $f^0_1 : M^0_1 \to \mathbf{R}^3$ can be constructed by using a 2-parameter deformation of Morse functions $T^2_{(3)} \to \mathbf{R}$ as depicted in Fig. 6.3, which corresponds to raising/lowering the three critical points.

Let B^1_j be a 2-disk fiber of the bundle $N(A^1_j) \to \widehat{A}^1_j$, $j = 1, 2, 3$. Then we can construct a compact orientable 3-manifold N^1_j and a smooth map $g^1_j : N^1_j \to B^1_j$ which has only fold points as its singularities such that its fibers are as depicted in Fig. 6.4 (for details, see [28, 30, 42], for example). Then we can construct a smooth map $f^1_j : M^1_j = N^1_j \times [0,1] \to N(A^1_j)$ by putting $f^1_j = g^1_j \times \mathrm{id}_{[0,1]}$, where we identify $N(A^1_j)$ with $B^1_j \times [0,1]$. Note that M^1_j is diffeomorphic to $T^2_{(2)} \times [-1,1] \times [0,1]$.

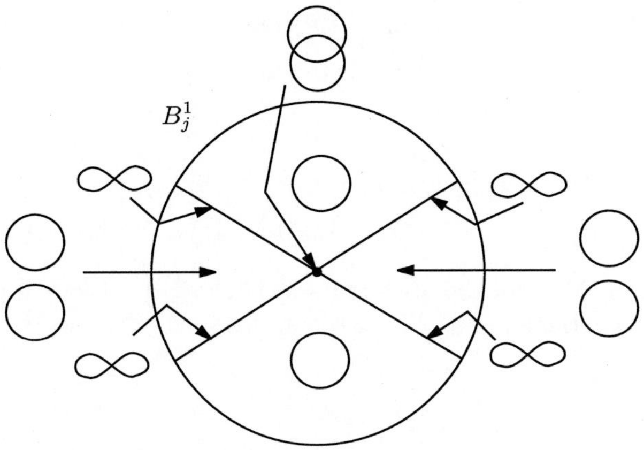

Fig. 6.4. Fibers over the points in B_j^1 for g_j^1

Similarly, for each of the four strata A_j^2 diffeomorphic to an open disk, $j = 1, 2, 3, 4$, by using a Morse function $S^2_{(3)} \to [-1, 1]$ as in Fig. 2.2 (2), we can construct a smooth map $f_j^2 : M_j^2 \to N(A_j^2) \cong [-1, 1] \times D^2$ which has only fold points as its singularities such that its fibers are as depicted in Fig. 6.2. Note that M_j^2 is diffeomorphic to $S^2_{(3)} \times D^2$. For the other three strata A_j^2 diffeomorphic to a 2-sphere, $j = 5, 6, 7$, we do not construct f_j^2 for the moment.

Now let us piece together the smooth maps constructed above. First, we attach $f_1^0 : M_1^0 \to N(A_1^0)$ and $f_j^1 : N_j^1 \times [0, 1] \to N(A_j^1)$, $j = 1, 2, 3$, by using appropriate embeddings $\varphi_j^1 : N_j^1 \times \{0, 1\} \to \partial M_1^0$. This is possible by the classification of singular fibers of C^∞ stable maps of 3-manifolds into surfaces (see Remark 3.14 and Theorem 3.15), since f_1^0 and f_j^1 have the same singular fiber of $\kappa = 2$ on the attaching part. Note that then the natural map

$$(f_1^0 \cup f_j^1)^{-1}((N(A_1^0) \cup N(A_j^1)) \cap N(A_j^2)) \to \partial \widehat{A}_j^2 \tag{6.1}$$

is the projection of a smooth $S^2_{(3)}$-bundle over a circle, $j = 1, 2, 3$.

Note that we have a nontrivial diffeomorphism $\varphi : N_j^1 \to N_j^1$ such that $g_j^1 \circ \varphi = g_j^1$. (This corresponds to the rotation through the angle π around the center of the square representing $T^2_{(2)}$ in [42, Fig. 1].) Thus, we may assume that the $S^2_{(3)}$-bundle (6.1) is trivial by changing the embedding φ_j^1 by $\varphi_j^1 \circ \widetilde{\varphi}$ if necessary, where $\widetilde{\varphi} : N_j^1 \times \{0, 1\} \to N_j^1 \times \{0, 1\}$ is the identity on $N_j^1 \times \{0\}$ and is φ on $N_j^1 \times \{1\}$. Let us denote the resulting map $f_1^0 \cup f_1^1 \cup f_2^1 \cup f_3^1$ by $\widetilde{f}^1$. Then, we can check that the natural map

$$(\widetilde{f}^1)^{-1}((N(A_1^0) \cup N(A_1^1) \cup N(A_2^1) \cup N(A_3^1)) \cap N(A_4^2)) \to \partial \widehat{A}_4^2 \tag{6.2}$$

is also the projection of a trivial $S^2_{(3)}$-bundle over a circle.

Since the $S^2_{(3)}$-bundles (6.1) and (6.2) are trivial, we can now attach $f_j^2 : M_j^2 \cong S^2_{(3)} \times D^2 \to N(A_j^2)$, $j = 1, 2, 3, 4$, to $\widetilde{f}^1$. Let us denote the resulting map $\widetilde{f}^1 \cup f_1^2 \cup f_2^2 \cup f_3^2 \cup f_4^2$ by

$$\widetilde{f}^2 : \widetilde{M}^2 \to N(A_1^0) \cup N(A_1^1) \cup N(A_2^1) \cup N(A_3^1) \\ \cup N(A_1^2) \cup N(A_2^2) \cup N(A_3^2) \cup N(A_4^2) \subset \mathbf{R}^3.$$

Note that the image X of the above map is nothing but the regular neighborhood of the Boy surface P. Let $\partial X = \partial_0 X \cup \partial_1 X$ be the connected components of ∂X, where

$$\partial_0 X = X \cap N(A_2^3) \quad \text{and} \quad \partial_1 X = X \cap N(A_3^3),$$

both of which are diffeomorphic to the 2-sphere. Note that

$$\widetilde{f}^2|_{(\widetilde{f}^2)^{-1}(\partial_0 X)} : (\widetilde{f}^2)^{-1}(\partial_0 X) \to \partial_0 X \tag{6.3}$$

is the projection of a smooth orientable S^1-bundle over a 2-sphere, and that

$$\widetilde{f}^2|_{(\widetilde{f}^2)^{-1}(\partial_1 X)} : (\widetilde{f}^2)^{-1}(\partial_1 X) \to \partial_1 X \tag{6.4}$$

is the projection of a smooth orientable $(S^1 \cup S^1)$-bundle over a 2-sphere. Note also that the latter is a disjoint union of two orientable S^1-bundles, since $\partial_1 X$ is simply connected.

Let M_5^2 be the total space of the D^2-bundle associated with the S^1-bundle (6.3), and M_6^2, M_7^2 the total spaces of the D^2-bundles associated with the two S^1-bundles (6.4). Then, by extending the maps (6.3) and (6.4), we can construct smooth maps

$$f_5^2 : M_5^2 \to N(A_5^2) \cup N(A_2^3), \tag{6.5}$$
$$f_6^2 : M_6^2 \to N(A_6^2) \cup N(A_3^3), \tag{6.6}$$
$$f_7^2 : M_7^2 \to N(A_7^2) \cup N(A_4^3) \cup N(A_6^2) \cup N(A_3^3) \tag{6.7}$$

with only definite fold points as their singularities such that their singular sets correspond to the zero sections of the D^2-bundles, $f_5^2(S_0(f_5^2)) = A_5^2$, $f_6^2(S_0(f_6^2)) = A_6^2$, and $f_7^2(S_0(f_7^2)) = A_7^2$. Then, their fibers are as depicted in Fig. 6.2. By our construction, we can glue (6.5), (6.6), (6.7) and $\widetilde{f}^2$ to get a smooth map

$$f : M \to \mathbf{R}^3$$

of a smooth closed 4-manifold M into $\mathbf{R}^3$.

Note that f has only fold points as its singularities and that its fibers are exactly as depicted in Fig. 6.2. Then by Proposition 3.1, f is a C^∞ stable map.

In order to prove that M is diffeomorphic to $\mathbf{C}P^2 \sharp 2\overline{\mathbf{C}P^2}$, let us consider a C^∞ stable map $g : M' \to \mathbf{R}^3$ constructed as follows. Let $Y = D_1^2 \cup (S^1 \times$

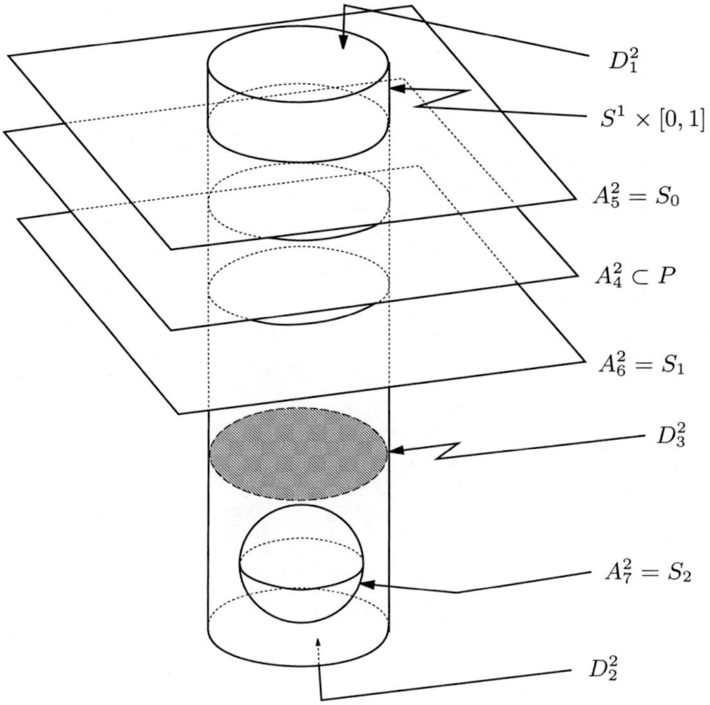

Fig. 6.5. Embedded 2-sphere $Y = D_1^2 \cup (S^1 \times [0,1]) \cup D_2^2$ in $\mathbf{R}^3$

$[0,1]) \cup D_2^2$ be a 2-sphere embedded in $\mathbf{R}^3$ which intersects A_5^2, A_4^2 and A_6^2 transversely as shown in Fig. 6.5, where D_1^2 and D_2^2 are copies of 2-disks. We take Y so that the 3-disk $\widetilde{Y}$ bounded by Y contains $A_7^2 = S_2$ in its interior. Note that the natural map

$$f^{-1}(S^1 \times [0,1]) \xrightarrow{\ f\ } S^1 \times [0,1] \xrightarrow{\ \pi_1\ } S^1$$

is a trivial D^2-bundle, where π_1 is the projection to the first factor. Note also that the map $h_x = f|_{f^{-1}(\{x\}\times[0,1])} : D^2 \to [0,1]$ is a Morse function as described in Fig. 6.6 for all $x \in S^1$ and is independent of the choice of x.

Let us replace the map $f|_{f^{-1}(\widetilde{Y})}$ by the smooth map $g_{\widetilde{Y}}$ whose fibers are as described in Fig. 6.7 (in fact, the real figure is obtained by rotating the rectangle around the vertical line in the center).

Let us explain the reason why such a replacement is possible. We identify $\widetilde{Y}$ with $D^2 \times [0,1]$ so that $D^2 \times \{\varepsilon\}$ corresponds to $D_{2-\varepsilon}^2$ for $\varepsilon = 0, 1$. Let Δ be a small concentric 2-disk in the interior of D^2. By using a generic deformation of functions $k_t : D^2 \to [0,1]$, $t \in [1/2, 1]$, as shown in Fig. 6.8, we can construct the smooth map

$$g_1 : S^1 \times [1/2,1] \times D^2 \to \overline{\widetilde{Y} \smallsetminus (\Delta \times [0,1])} \cong S^1 \times [1/2,1] \times [0,1]$$

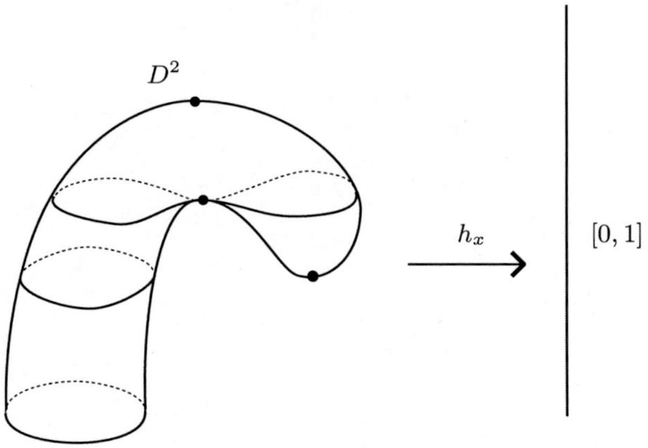

Fig. 6.6. Morse function $h_x : D^2 \to [0,1]$

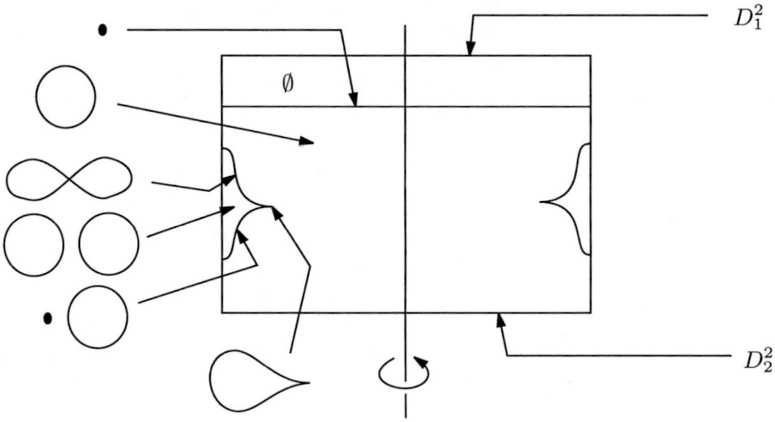

Fig. 6.7. Fibers of $g_{\widetilde{Y}}$

by putting $g_1(x,t,q) = (x,t,k_t(q))$. Note that g_1 has only fold points and cusp points as its singularities and is consistent with $f|_{f^{-1}(\mathbf{R}^3 \smallsetminus \mathrm{Int}\, \widetilde{Y})}$ along $(S^1 \times \{1\} \times D^2) \cup (S^1 \times [1/2,1] \times \partial D^2)$.

Then, using the Morse function $k_{1/2}$, we define the smooth map $g_2 : \Delta \times D^2 \to \Delta \times [0,1]$ by $g_2(x,q) = (x, k_{1/2}(q))$. Obviously, this is consistent with $g_1|_{g_1^{-1}(\partial\Delta \times [0,1])}$ along $\partial\Delta \times D^2 = S^1 \times \{1/2\} \times D^2$, although we do not know if it is consistent with $f|_{f^{-1}(\mathbf{R}^3 \smallsetminus \mathrm{Int}\, \widetilde{Y})}$ along

$$g_2^{-1}(\Delta \times \{0\}) = \Delta \times \partial D^2 = f^{-1}(\Delta \times \{0\}). \tag{6.8}$$

However, we have a plenty of diffeomorphisms $D^2 \to D^2$ that preserve the Morse function $k_{1/2}$. For example, all the diffeomorphisms in the rotation group $SO(2)$ satisfy this property. Hence, changing the identification

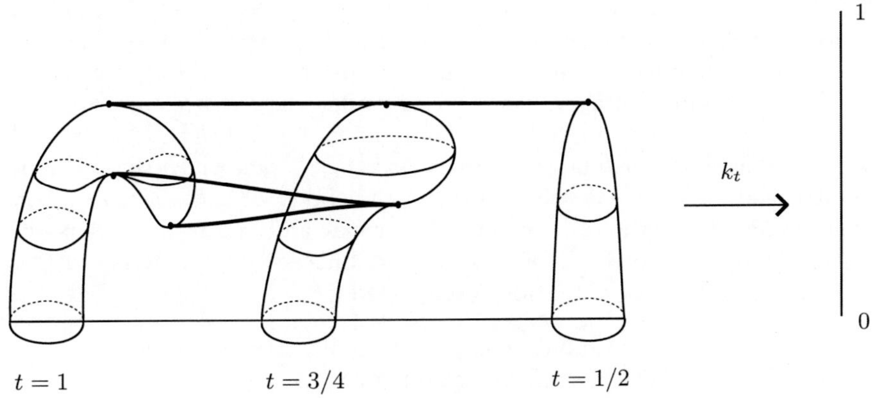

Fig. 6.8. Deformation of functions on the 2-disk

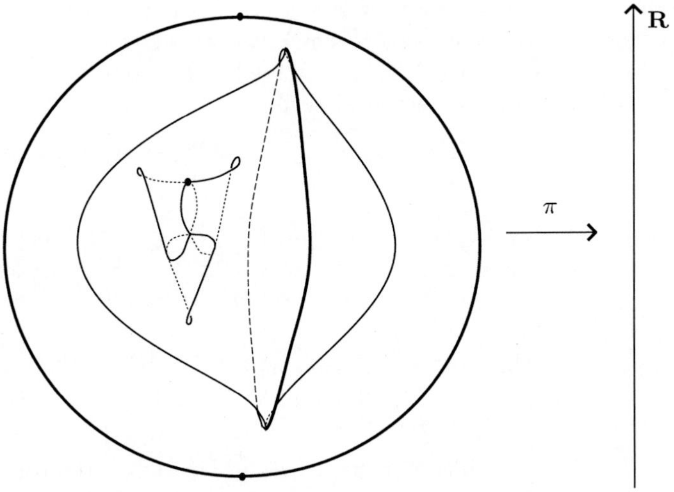

Fig. 6.9. Image of $S(g)$ by g

$g_2^{-1}(\partial\Delta \times [0, 1]) \cong \partial\Delta \times D^2$ if necessary, we can arrange so that g_2 is consistent with $f|_{f^{-1}(\mathbf{R}^3 \setminus \mathrm{Int}\,\widetilde{Y})}$ along (6.8). Therefore, we obtain a C^∞ stable map $g : M' \to \mathbf{R}^3$ by gluing $f|_{f^{-1}(\mathbf{R}^3 \setminus \mathrm{Int}\,\widetilde{Y})}$, g_1 and g_2, where $g_{\widetilde{Y}} = g_1 \cup g_2$ (see Fig. 6.7 again). Note that the singular set $S(g)$ is the union of a 2-sphere component consisting of definite fold points and a projective plane component containing the set of cusp points.

Lemma 6.1. *The smooth closed 4-manifold M' is diffeomorphic to $\mathbf{C}P^2$ or $\overline{\mathbf{C}P^2}$.*

Proof. Let $\pi : \mathbf{R}^3 \to \mathbf{R}$ be a projection. By locating $g(S(g))$ as in Fig. 6.9 by an isotopy of $\mathbf{R}^3$, we may assume that $\pi \circ g : M' \to \mathbf{R}$ is a Morse function with exactly three critical points (for such a construction of Morse functions, refer to [14] for more details). We see easily that their indices are equal to 0, 2 and 4. Thus, M' has a handlebody decomposition $h^0 \cup h^2 \cup h^4$, where h^i denotes an i-handle. Let k be the knot in $\partial h^0 = S^3$ along which the 2-handle h^2 is attached to h^0. Since the resulting handlebody $h^0 \cup h^2$ has boundary diffeomorphic to S^3, the knot k must be trivial and the framing must be equal to ± 1 by a deep result of Gordon and Luecke [17, 18]. Hence, M' is diffeomorphic to $\mathbf{C}P^2$ or $\overline{\mathbf{C}P^2}$ (for details, see [25]).

Remark 6.2. In this way we have completed the construction of a C^∞ stable map $g : \mathbf{C}P^2 \to \mathbf{R}^3$ with the following properties.

(1) The map g has only fold and cusp points as its singularities.
(2) The set $C(g)$ of its cusp points constitutes a circle, and the singular set $S(g)$ is the union of a 2-sphere component consisting of definite fold points and a projective plane component which contains $C(g)$.
(3) The discriminant set $g(S(g))$ is as described in Fig. 6.9.
(4) The fibers of g can be completely described.

Presumably, the C^∞ stable map $g : M' = \mathbf{C}P^2 \to \mathbf{R}^3$ thus constructed coincides with Kobayashi's example presented in [26, 27].

By choosing an appropriate orientation for M', we may assume that it is orientation preservingly diffeomorphic to $\overline{\mathbf{C}P^2}$. By our construction, it is easy to see that $g^{-1}(\widetilde{Y})$ is diffeomorphic to D^4. Hence $f^{-1}(\mathbf{R}^3 \setminus \operatorname{Int} \widetilde{Y})$ is diffeomorphic to $\overline{\mathbf{C}P^2} - \operatorname{Int} D^4$.

Let us determine the diffeomorphism type of $f^{-1}(\widetilde{Y})$. Take a properly embedded 2-disk D^2_3 in $\widetilde{Y}$ as in Fig. 6.5, and let $\widetilde{Y}_1$ and $\widetilde{Y}_2$ be the 3-disks such that $\widetilde{Y} = \widetilde{Y}_1 \cup \widetilde{Y}_2$, $\widetilde{Y}_1 \cap \widetilde{Y}_2 = D^2_3$, and $\widetilde{Y}_2 \supset S_2$. Then it is easy to see that $f^{-1}(\widetilde{Y}_1)$ and $f^{-1}(\widetilde{Y}_2)$ are diffeomorphic to $D^2 \times D^2$ and to the total space E of a D^2-bundle over S^2, respectively. More precisely, $f^{-1}(\widetilde{Y})$ is obtained from E by attaching a 2-handle along the boundary of a D^2-fiber of the fibration $E \to S^2$. Hence, $f^{-1}(\widetilde{Y})$ is diffeomorphic either to $\mathbf{C}P^2 \sharp \overline{\mathbf{C}P^2} \setminus \operatorname{Int} D^4$ or to $S^2 \times S^2 \setminus \operatorname{Int} D^4$.

Therefore, the source 4-manifold $M = f^{-1}(\mathbf{R}^3)$ of f is diffeomorphic either to $\overline{\mathbf{C}P^2} \sharp (\mathbf{C}P^2 \sharp \overline{\mathbf{C}P^2})$ or to $\overline{\mathbf{C}P^2} \sharp (S^2 \times S^2)$. In both cases, M is diffeomorphic to $\mathbf{C}P^2 \sharp 2\overline{\mathbf{C}P^2}$ (for details, see [25], for example). This completes the construction of the desired C^∞ stable map $f : \mathbf{C}P^2 \sharp 2\overline{\mathbf{C}P^2} \to \mathbf{R}^3$ as promised at the beginning of this chapter.

It is an easy task to check that all the results obtained in Chaps. 4 and 5 are valid for the above constructed C^∞ stable maps.

Remark 6.3. The author has shown that $\mathbf{C}P^2$ does not admit a fold map into $\mathbf{R}^3$ (see [42, 45, 53, 1, 47, 39]). This implies that the normal bundle of the definite fold component of f in M corresponding to S_2 is nontrivial, for if it were trivial, then we could construct a smooth map $g'' : M' \to \mathbf{R}^3$ with only fold points as its singularities. In fact, we can show that the normal Euler numbers of the definite fold components of f in M corresponding to S_0, S_1 and S_2 are equal to $1, -2$ and -2 respectively (for details, see [50]).

Using the example constructed above, we can show the following.

Proposition 6.4. *For every $n \geq 1$, there exists a smooth map*

$$f_n : n\mathbf{C}P^2 \sharp (n+1)\overline{\mathbf{C}P^2} \to \mathbf{R}^3$$

with only fold points as its singularities.

Proof. Recall that there exists a smooth map $\ell : \mathbf{C}P^2 \sharp \overline{\mathbf{C}P^2} \to \mathbf{R}^3$ with only definite fold points as its singularities (for example, see [43]). Note also that such a map can be constructed explicitly. Then, we can construct the desired map f_n from $f = f_1 : \mathbf{C}P^2 \sharp 2\overline{\mathbf{C}P^2} \to \mathbf{R}^3$ and $n-1$ copies of ℓ by the connected sum construction (for details, see [43]).

Remark 6.5. Sakuma [52] had conjectured that no closed orientable 4-manifold with odd Euler characteristic can admit a fold map into $\mathbf{R}^3$ (see also [24, Remark 2.3]). The above proposition gives explicit counter-examples to his conjecture. Note that a more precise result has been obtained in [47] about fold maps of 4-manifolds into $\mathbf{R}^3$ (see also [41]).

Part II

Universal Complex of Singular Fibers

7

Generalities

In this chapter, we begin to formalize the idea used in Chaps. 4 and 5.

First, let us prepare the following notation. For a pair of nonnegative integers (n,p), we denote by $\mathcal{T}_{\mathrm{pr}}(n,p)$ (or by $\mathcal{S}^{\infty}_{\mathrm{pr}}(n,p)$) the set of all proper Thom maps (resp. proper C^∞ stable maps) between manifolds of dimensions n and p (for Thom maps, see Example 1.3 of Chap. 1). Furthermore, we denote by $\mathcal{S}^0_{\mathrm{pr}}(n,p)$ the set of all C^0 stable maps which are elements of $\mathcal{T}_{\mathrm{pr}}(n,p)$. Note that we have $\mathcal{S}^{\infty}_{\mathrm{pr}}(n,p) \subset \mathcal{T}_{\mathrm{pr}}(n,p)$. However, the author does not know if a proper C^0 stable map is a Thom map or not, so that we adopt the above convention. Note also that $\mathcal{S}^0_{\mathrm{pr}}(n,p) = \mathcal{S}^{\infty}_{\mathrm{pr}}(n,p)$ for nice dimension pairs (n,p) in the sense of Mather [32] by [12, 60] (see also Remark 3.2).

In the following, we call $k = p - n$ the *codimension* of a map in these sets. For a fixed k, we put

$$\widetilde{\mathcal{T}}_{\mathrm{pr}}(k) = \bigcup_{p-n=k} \mathcal{T}_{\mathrm{pr}}(n,p),$$

$$\widetilde{\mathcal{S}}^{\infty}_{\mathrm{pr}}(k) = \bigcup_{p-n=k} \mathcal{S}^{\infty}_{\mathrm{pr}}(n,p),$$

$$\widetilde{\mathcal{S}}^{0}_{\mathrm{pr}}(k) = \bigcup_{p-n=k} \mathcal{S}^{0}_{\mathrm{pr}}(n,p).$$

In the following, for a Thom map $f : M \to N$ in $\mathcal{T}_{\mathrm{pr}}(n,p)$, $\mathcal{M}$ and $\mathcal{N}$ will denote Whitney stratifications of M and N respectively such that f satisfies the Thom regularity condition [15, Chapter I, §3] with respect to them. For a C^0 equivalence class $\mathfrak{F}$ of fibers, we denote by $\mathfrak{F}(f)$ the set of points in N over which lies a fiber of type $\mathfrak{F}$.

Lemma 7.1. *The subspace $\mathfrak{F}(f)$ of N is a union of strata of $\mathcal{N}$ and is a C^0 submanifold of N of constant codimension if it is nonempty. Furthermore, this codimension does not depend on a particular choice of $f \in \mathcal{T}_{\mathrm{pr}}(n,p)$.*

Proof. The first assertion has already been shown in Example 1.3. In order to show the second assertion, let us take a top dimensional stratum Σ contained in $\mathfrak{F}(f)$. Note that for each point $y \in \Sigma$, there exists a neighborhood U_y of y in N such that $U_y \cap \mathfrak{F}(f) = U_y \cap \Sigma$, since Σ is top dimensional. On the other hand, by the definition of C^0 equivalence, for each point y' of $\mathfrak{F}(f)$, there exists a neighborhood $U_{y'}$ of y' in N such that $(U_{y'}, U_{y'} \cap \mathfrak{F}(f))$ is homeomorphic to $(U_y, U_y \cap \mathfrak{F}(f))$. Hence the assertion about $\mathfrak{F}(f)$ follows. Using a similar argument, we can prove the final assertion. This completes the proof. □

Note that by virtue of the above lemma, the codimension of a C^0 type $\mathfrak{F}$ of fibers makes sense, and we denote it by $\kappa(\mathfrak{F})$.

Let us introduce the following notion which will play an important role throughout the rest of the book.

Definition 7.2. Suppose that an equivalence relation $\varrho = \varrho_{n,p}$ among the fibers of proper Thom maps between smooth manifolds of dimensions n and p is given. We say that the relation ϱ is *admissible* if the following conditions are satisfied.

(1) If two fibers are C^0 equivalent, then they are also equivalent with respect to ϱ.
(2) For any two proper Thom maps $f_i : M_i \to N_i$ in $\mathcal{T}_{\mathrm{pr}}(n, p)$ and for any points $y_i \in N_i$, $i = 0, 1$, such that the fibers over y_i are equivalent to each other with respect to ϱ, there exist neighborhoods U_i of y_i in N_i, $i = 0, 1$, and a homeomorphism $\varphi : U_0 \to U_1$ such that $\varphi(y_0) = y_1$ and $\varphi(U_0 \cap \widetilde{\mathfrak{F}}(f_0)) = U_1 \cap \widetilde{\mathfrak{F}}(f_1)$ for every equivalence class $\widetilde{\mathfrak{F}}$ of fibers with respect to ϱ, where $\widetilde{\mathfrak{F}}(f_i)$ is the set of points in N_i over which lies a fiber of f_i of type $\widetilde{\mathfrak{F}}$.

For example, the C^0 equivalence is clearly admissible in the above sense. We denote the C^0 equivalence relation among the fibers of elements of $\mathcal{T}_{\mathrm{pr}}(n, p)$ by $\varrho^0_{n,p}$.

In the following argument, we fix an admissible equivalence relation $\varrho = \varrho_{n,p}$ as in Definition 7.2.

Lemma 7.3. *For every equivalence class $\widetilde{\mathfrak{F}}$ with respect to an admissible equivalence relation ϱ, and for every proper Thom map $f : M \to N$ in $\mathcal{T}_{\mathrm{pr}}(n, p)$, the subspace $\widetilde{\mathfrak{F}}(f)$ of N is a union of strata of $\mathcal{N}$ and is a C^0 submanifold of N of constant codimension if it is nonempty. Furthermore, this codimension does not depend on a particular choice of $f \in \mathcal{T}_{\mathrm{pr}}(n, p)$.*

Proof. By Definition 7.2 (1) and Lemma 7.1, $\widetilde{\mathfrak{F}}(f)$ is a union of strata. Hence, the rest of the assertion follows from an argument similar to that in the proof of Lemma 7.1 together with Definition 7.2 (2). □

By virtue of the above lemma, the codimension of $\widetilde{\mathfrak{F}}$ makes sense, and we denote it by $\kappa(\widetilde{\mathfrak{F}})$.

For an equivalence class $\widetilde{\mathfrak{F}}$ of fibers with respect to ϱ with $\kappa = \kappa(\widetilde{\mathfrak{F}})$, let $\partial\widetilde{\mathfrak{F}}$ be the set of equivalence classes $\widetilde{\mathfrak{G}}$ of fibers with respect to ϱ of codimension

$\kappa + 1$ such that $\widetilde{\mathfrak{G}}(f) \subset \overline{\widetilde{\mathfrak{F}}(f)} \setminus \widetilde{\mathfrak{F}}(f)$ for every $f \in \mathcal{T}_{\mathrm{pr}}(n,p)$. For $\widetilde{\mathfrak{G}} \in \partial\widetilde{\mathfrak{F}}$, we take a proper Thom map $f \in \mathcal{T}_{\mathrm{pr}}(n,p)$ with $\widetilde{\mathfrak{G}}(f) \neq \emptyset$. Then we take a top dimensional stratum $\Sigma \subset \widetilde{\mathfrak{G}}(f)$, and let B_Σ be a small disk which intersects Σ transversely exactly at its center and whose dimension coincides with the codimension of Σ. Then $B_\Sigma \cap \overline{\widetilde{\mathfrak{F}}(f)}$ consists of a finite number of arcs which have $B_\Sigma \cap \Sigma$ as a common end point. Let $n_{\widetilde{\mathfrak{F}}}(\widetilde{\mathfrak{G}}) \in \mathbf{Z}_2$ denote the number of such arcs modulo two, which clearly does not depend on the choice of B_Σ, Σ or f by Definition 7.2 (2). Then, by considering the homological boundary of $\overline{\widetilde{\mathfrak{F}}(f)}$, we have the following.

Proposition 7.4. *For every equivalence class $\widetilde{\mathfrak{F}}$ of fibers with respect to an admissible equivalence relation ϱ, and for every $f : M \to N$ in $\mathcal{T}_{\mathrm{pr}}(n,p)$, the $\mathbf{Z}_2$-chain*

$$\sum_{\widetilde{\mathfrak{G}} \in \partial\widetilde{\mathfrak{F}}} n_{\widetilde{\mathfrak{F}}}(\widetilde{\mathfrak{G}})\overline{\widetilde{\mathfrak{G}}(f)} \tag{7.1}$$

(of closed support) is a cycle in N and represents the zero homology class in the homology $H^c_{p-\kappa-1}(N;\mathbf{Z}_2)$ of closed support, where κ denotes the codimension of $\widetilde{\mathfrak{F}}$.

Proof. By the definition of $n_{\widetilde{\mathfrak{F}}}(\widetilde{\mathfrak{G}})$, we see that the $\mathbf{Z}_2$-chain (7.1) coincides the boundary of the $\mathbf{Z}_2$-chain $\overline{\widetilde{\mathfrak{F}}(f)}$ in N. Hence the result follows. □

Remark 7.5. In the above proposition, if $\widetilde{\mathfrak{F}}$ does not contain the empty fiber and the source manifold M is compact, then the $\mathbf{Z}_2$-chain (7.1) has compact support and represents the zero homology class in the usual homology $H_{p-\kappa-1}(N;\mathbf{Z}_2)$.

We warn the reader that the sum appearing in the right hand side of (7.1) may contain infinitely many terms if the source manifold M of f is not compact.

Note that all the results obtained in Chap. 4 are special cases of the above proposition. Some applications of Proposition 7.4 to other specific situations will be given in Chap. 15.

8

Universal Complex of Singular Fibers

In this chapter, based on the idea given in the previous chapter, we define a complex of singular fibers for a specific map, and then we define its universal versions for various classes of maps. We will see later that this is a generalization of Vassiliev's universal complex of multi-singularities [58]. Here we develop a rather detailed theory of such universal complexes in order to better understand what is the essential point behind our results obtained in Chaps. 4 and 5, and to obtain further related results.

8.1 Complex of Singular Fibers for a Specific Map

Let $f : M \to N$ be a proper smooth map of a smooth n-dimensional manifold M into a smooth p-dimensional manifold N such that f is a Thom map in the sense of Example 1.3, as in the previous chapter: in other words, $f \in \mathcal{T}_{\mathrm{pr}}(n,p)$.

In the following, we fix an equivalence relation $\varrho = \varrho_{n,p}$ for the set of fibers of such maps which is admissible in the sense of Definition 7.2. Let us construct a complex of fibers for f with coefficients in $\mathbf{Z}_2$ with respect to the admissible equivalence relation ϱ as follows.

For $\kappa \geq 0$, let $C^\kappa(f,\varrho)$ be the $\mathbf{Z}_2$-vector space consisting of all formal linear combinations,

$$\sum_{\kappa(\widetilde{\mathfrak{F}})=\kappa} m_{\widetilde{\mathfrak{F}}}\widetilde{\mathfrak{F}} \quad (m_{\widetilde{\mathfrak{F}}} \in \mathbf{Z}_2),$$

which may possibly contain infinitely many terms if M is noncompact, of the equivalence classes $\widetilde{\mathfrak{F}}$ of fibers of f with codimension κ with respect to the equivalence relation ϱ. If there are no such fibers, then we simply put $C^\kappa(f,\varrho) = 0$. Furthermore, for $\kappa < 0$, we also put $C^\kappa(f,\varrho) = 0$. For two equivalence classes of fibers $\widetilde{\mathfrak{F}}$ and $\widetilde{\mathfrak{G}}$ of f with $\kappa(\widetilde{\mathfrak{F}}) = \kappa(\widetilde{\mathfrak{G}}) - 1$, we define the *incidence coefficient* $[\widetilde{\mathfrak{F}} : \widetilde{\mathfrak{G}}]_f \in \mathbf{Z}_2$ by putting $[\widetilde{\mathfrak{F}} : \widetilde{\mathfrak{G}}]_f = n_{\widetilde{\mathfrak{F}}}(\widetilde{\mathfrak{G}}) \in \mathbf{Z}_2$ if $\widetilde{\mathfrak{G}}(f) \subset \overline{\widetilde{\mathfrak{F}}(f)} \setminus \widetilde{\mathfrak{F}}(f)$, and $[\widetilde{\mathfrak{F}} : \widetilde{\mathfrak{G}}]_f = 0$ otherwise. Define the $\mathbf{Z}_2$-linear map

$$\delta_\kappa(f) : C^\kappa(f, \varrho) \to C^{\kappa+1}(f, \varrho)$$

by

$$\delta_\kappa(f)(\widetilde{\mathfrak{F}}) = \sum_{\kappa(\widetilde{\mathfrak{G}})=\kappa+1} [\widetilde{\mathfrak{F}} : \widetilde{\mathfrak{G}}]_f \widetilde{\mathfrak{G}}, \tag{8.1}$$

for $\widetilde{\mathfrak{F}}$ with $\kappa(\widetilde{\mathfrak{F}}) = \kappa$. We warn the reader that the sum appearing in the right hand side of (8.1) may possibly contain infinitely many terms if M is noncompact. Nevertheless, for a given equivalence class $\widetilde{\mathfrak{G}}$ of fibers of f with codimension $\kappa + 1$, the number of equivalence classes $\widetilde{\mathfrak{F}}$ of fibers of f with codimension κ such that $[\widetilde{\mathfrak{F}} : \widetilde{\mathfrak{G}}]_f \neq 0$ is finite by virtue of the local finiteness of the Whitney regular stratifications and the definition of an admissible equivalence relation. Hence, the linear map $\delta_\kappa(f)$ is well-defined.

The following lemma can be proved by an argument similar to that in [58, 8.3.4 Lemma] or [38, Lemma 1.5]. Details are left to the reader.

Lemma 8.1. $\delta_{\kappa+1}(f) \circ \delta_\kappa(f) = 0$.

Therefore, $\mathcal{C}(f, \varrho) = (C^\kappa(f, \varrho), \delta_\kappa(f))_\kappa$ constitutes a complex and its cohomology groups $H^\kappa(f, \varrho)$ are well-defined.

8.2 Complex for Maps Between Manifolds of Fixed Dimensions

The above construction can be generalized to get a "universal" complex of singular fibers for proper Thom maps between manifolds of dimensions n and p as follows.

Let ϱ be an admissible equivalence relation as in Definition 7.2 for the fibers of elements of $\mathcal{T}_{\mathrm{pr}}(n, p)$. For $\kappa \in \mathbf{Z}$, let $C^\kappa(\mathcal{T}_{\mathrm{pr}}(n, p), \varrho)$ be the $\mathbf{Z}_2$-vector space consisting of all formal linear combinations,

$$\sum_{\kappa(\widetilde{\mathfrak{F}})=\kappa} m_{\widetilde{\mathfrak{F}}} \widetilde{\mathfrak{F}} \quad (m_{\widetilde{\mathfrak{F}}} \in \mathbf{Z}_2),$$

which may possibly contain infinitely many terms, of the equivalence classes $\widetilde{\mathfrak{F}}$ of fibers of proper Thom maps between manifolds of dimensions n and p with $\kappa(\widetilde{\mathfrak{F}}) = \kappa$ with respect to the equivalence relation $\varrho = \varrho_{n,p}$. If there is no such equivalence class (for example, if $\kappa > p$ or $\kappa < 0$), then we simply put $C^\kappa(\mathcal{T}_{\mathrm{pr}}(n, p), \varrho) = 0$. For two equivalence classes $\widetilde{\mathfrak{F}}$ and $\widetilde{\mathfrak{G}}$ of fibers of elements of $\mathcal{T}_{\mathrm{pr}}(n, p)$ with $\kappa(\widetilde{\mathfrak{F}}) = \kappa(\widetilde{\mathfrak{G}}) - 1$, we define the *incidence coefficient* $[\widetilde{\mathfrak{F}} : \widetilde{\mathfrak{G}}] \in \mathbf{Z}_2$ by putting $[\widetilde{\mathfrak{F}} : \widetilde{\mathfrak{G}}] = n_{\widetilde{\mathfrak{F}}}(\widetilde{\mathfrak{G}}) \in \mathbf{Z}_2$ if $\widetilde{\mathfrak{G}}(f) \subset \overline{\widetilde{\mathfrak{F}}(f)} \setminus \widetilde{\mathfrak{F}}(f)$ for every $f \in \mathcal{T}_{\mathrm{pr}}(n, p)$, and $[\widetilde{\mathfrak{F}} : \widetilde{\mathfrak{G}}] = 0$ otherwise. Then the $\mathbf{Z}_2$-linear map $\delta_\kappa : C^\kappa(\mathcal{T}_{\mathrm{pr}}(n, p), \varrho) \to C^{\kappa+1}(\mathcal{T}_{\mathrm{pr}}(n, p), \varrho)$ is defined by

$$\delta_\kappa(\widetilde{\mathfrak{F}}) = \sum_{\kappa(\widetilde{\mathfrak{G}})=\kappa+1} [\widetilde{\mathfrak{F}} : \widetilde{\mathfrak{G}}]\widetilde{\mathfrak{G}}, \tag{8.2}$$

for $\widetilde{\mathfrak{F}}$ with $\kappa(\widetilde{\mathfrak{F}}) = \kappa$. (See (8.1) and the subsequent remark). Note that the incidence coefficient, and hence the map δ_κ, is well-defined by virtue of Definition 7.2 (2). Furthermore, we can prove that $\delta_{\kappa+1} \circ \delta_\kappa = 0$ as in Lemma 8.1. We call the resulting complex $\mathcal{C}(\mathcal{T}_{\mathrm{pr}}(n,p), \varrho) = (C^\kappa(\mathcal{T}_{\mathrm{pr}}(n,p), \varrho), \delta_\kappa)_\kappa$ the *universal complex of singular fibers for proper Thom maps between manifolds of dimensions n and p with respect to the admissible equivalence relation* $\varrho = \varrho_{n,p}$, and we denote its cohomology group of dimension κ by $H^\kappa(\mathcal{T}_{\mathrm{pr}}(n,p), \varrho)$.

For $f \in \mathcal{T}_{\mathrm{pr}}(n,p)$, let $C^\kappa(f^c, \varrho)$ be the linear subspace of $C^\kappa(\mathcal{T}_{\mathrm{pr}}(n,p), \varrho)$ spanned by those equivalence classes of fibers of elements of $\mathcal{T}_{\mathrm{pr}}(n,p)$ of codimension κ with respect to ϱ which contain no fiber of f.

Lemma 8.2. *For $f \in \mathcal{T}_{\mathrm{pr}}(n,p)$, the following holds.*

(1) *We have $\delta_\kappa(C^\kappa(f^c, \varrho)) \subset C^{\kappa+1}(f^c, \varrho)$ for every $\kappa \in \mathbf{Z}$. Hence, $\mathcal{C}(f^c, \varrho) = (C^\kappa(f^c, \varrho), \delta_\kappa|_{C^\kappa(f^c,\varrho)})_\kappa$ constitutes a subcomplex of $\mathcal{C}(\mathcal{T}_{\mathrm{pr}}(n,p), \varrho)$.*
(2) *The quotient complex*

$$\mathcal{C}(\mathcal{T}_{\mathrm{pr}}(n,p), \varrho)/\mathcal{C}(f^c, \varrho) = (C^\kappa(\mathcal{T}_{\mathrm{pr}}(n,p), \varrho)/C^\kappa(f^c, \varrho), \overline{\delta}_\kappa)_\kappa$$

is naturally isomorphic to $\mathcal{C}(f, \varrho)$, where

$$\overline{\delta}_\kappa : C^\kappa(\mathcal{T}_{\mathrm{pr}}(n,p), \varrho)/C^\kappa(f^c, \varrho) \to C^{\kappa+1}(\mathcal{T}_{\mathrm{pr}}(n,p), \varrho)/C^{\kappa+1}(f^c, \varrho)$$

is the well-defined $\mathbf{Z}_2$-linear map induced by δ_κ.

Proof. Let $\widetilde{\mathfrak{F}} \in C^\kappa(f^c, \varrho)$ be an equivalence class of fibers of codimension κ which contains no fiber of f. For an equivalence class $\widetilde{\mathfrak{G}} \in C^{\kappa+1}(\mathcal{T}_{\mathrm{pr}}(n,p), \varrho)$ of fibers of codimension $\kappa+1$, if $[\widetilde{\mathfrak{F}} : \widetilde{\mathfrak{G}}] \neq 0$, then $\widetilde{\mathfrak{G}}(f) \subset \overline{\widetilde{\mathfrak{F}}(f)} \setminus \widetilde{\mathfrak{F}}(f)$. Since $\widetilde{\mathfrak{F}}$ does not contain any fiber of f, we have $\widetilde{\mathfrak{F}}(f) = \emptyset$, and hence $\widetilde{\mathfrak{G}}(f) = \emptyset$. Thus, we have $\widetilde{\mathfrak{G}} \in C^{\kappa+1}(f^c, \varrho)$ and item (1) follows.

Let $\pi_\kappa : C^\kappa(\mathcal{T}_{\mathrm{pr}}(n,p), \varrho) \to C^\kappa(f, \varrho)$ be the natural projection: i.e., π_κ is the linear map defined by

$$\pi_\kappa(\widetilde{\mathfrak{F}}) = \begin{cases} \widetilde{\mathfrak{F}}, & \text{if } \widetilde{\mathfrak{F}} \in C^\kappa(f, \varrho), \\ 0, & \text{otherwise}, \end{cases}$$

for an equivalence class $\widetilde{\mathfrak{F}} \in C^\kappa(\mathcal{T}_{\mathrm{pr}}(n,p), \varrho)$ of fibers. Then, it is easy to see that the system of $\mathbf{Z}_2$-linear maps $\{\pi_\kappa\}_\kappa$ defines a surjective cochain map and the kernel of π_κ coincides with $C^\kappa(f^c, \varrho)$. Hence, item (2) follows. This completes the proof. □

In view of the above lemma, the complex $\mathcal{C}(\mathcal{T}_{\mathrm{pr}}(n,p), \varrho)$ is universal in the sense that the complex $\mathcal{C}(f, \varrho)$ for a specific Thom map f is obtained as a quotient complex.

Remark 8.3. We will see in Chap. 9 that the universal complex of singular fibers with respect to the C^0 equivalence as defined above corresponds to increasing the generators of each cochain group of Vassiliev's universal complex of multi-singularities [58] according to the topological structures of the fibers (see Definition 9.9 and Remark 9.10).

8.3 Complex for Maps with Fixed Codimension

As we have noticed in Remark 3.14, a singular fiber of a codimension k map into a p-dimensional manifold can naturally be identified with a singular fiber of a codimension k map into a $(p+1)$-dimensional manifold. This is formalized as follows.

Definition 8.4. Let $f : M \to N$ be a proper Thom map between manifolds of dimensions n and p with $k = p - n$. For a positive integer ℓ, we call the map

$$f \times \mathrm{id}_{\mathbf{R}^\ell} : M \times \mathbf{R}^\ell \to N \times \mathbf{R}^\ell$$

the *ℓ-th suspension* of f. (When $\ell = 1$, we sometimes call it the *suspension* of f and denote it by Σf.) Furthermore, to the fiber of f over a point $y \in N$, we can associate the fiber of $f \times \mathrm{id}_{\mathbf{R}^\ell}$ over $y \times \{0\}$. We say that the latter fiber is obtained from the original fiber by the *ℓ-th suspension.* Note that the ℓ-th suspension of a proper Thom map is again a proper Thom map. Note also that a fiber and its suspensions are all diffeomorphic to each other in the sense of Definition 1.1 (2).

By considering the suspension as above, we can define a cochain map

$$\mathcal{C}(\mathcal{T}_{\mathrm{pr}}(n+\ell, p+\ell), \varrho_{n+\ell,p+\ell}) \to \mathcal{C}(\mathcal{T}_{\mathrm{pr}}(n,p), \varrho_{n,p})$$

as long as the equivalence relations for the dimension pairs are consistent with each other in a certain sense, which is specified as follows.

Definition 8.5. Let us fix an integer k. Suppose that for each dimension pair (n, p) with $p - n = k$ and $\min(n,p) \geq 0$, we are given an admissible equivalence relation $\varrho_{n,p}$ for the fibers of proper Thom maps between manifolds of dimensions n and p. Such a system of equivalence relations

$$\mathcal{R}_k = \{\varrho_{n,p} \; : \; p - n = k, \, \min(n,p) \geq 0\},$$

which is often written simply as $\{\varrho_{n,p}\}_{p-n=k}$ or $\{\varrho_{p-k,p}\}_p$, is said to be *stable* if the following condition is satisfied: if two fibers of proper Thom maps between manifolds of dimensions n and p are equivalent with respect to $\varrho_{n,p}$, then their ℓ-th suspensions are also equivalent with respect to $\varrho_{n+\ell,p+\ell}$ for all $\ell > 0$. Note that the ℓ-th suspensions are fibers of proper Thom maps between manifolds of dimensions $n + \ell$ and $p + \ell$.

For example, the set of C^0 equivalence relations $\{\varrho^0_{p-k,p}\}_p$ gives a stable system of admissible equivalence relations for the fibers of proper Thom maps of codimension k, and we denote it by $\mathcal{R}^0_k$.

Suppose that a stable system of admissible equivalence relations $\mathcal{R}_k$ as in Definition 8.5 is given for the fibers of proper Thom maps of codimension k. Then, for every pair (n,p) with $p-n=k$ and a positive integer ℓ, the suspension induces a natural map

$$s_\kappa : C^\kappa(\mathcal{T}_{\rm pr}(n+\ell,p+\ell),\varrho_{n+\ell,p+\ell}) \to C^\kappa(\mathcal{T}_{\rm pr}(n,p),\varrho_{n,p}) \tag{8.3}$$

for $\kappa \in \mathbf{Z}$. More precisely, when $0 \le \kappa \le p$, for an equivalence class $\widetilde{\mathfrak{F}} \in C^\kappa(\mathcal{T}_{\rm pr}(n+\ell,p+\ell),\varrho_{n+\ell,p+\ell})$ of fibers with respect to $\varrho_{n+\ell,p+\ell}$, we define $s_\kappa(\widetilde{\mathfrak{F}}) \in C^\kappa(\mathcal{T}_{\rm pr}(n,p),\varrho_{n,p})$ to be the (possibly infinite) sum of all those equivalence classes of fibers of codimension κ with respect to $\varrho_{n,p}$ whose ℓ-th suspensions are contained in $\widetilde{\mathfrak{F}}$. For $\kappa > p$ or $\kappa < 0$, we simply put $s_\kappa = 0$. Note that s_κ is a well-defined $\mathbf{Z}_2$-linear map by virtue of Definition 8.5.

Lemma 8.6. *The $\mathbf{Z}_2$-linear map s_κ of (8.3) is a monomorphism for every $\kappa \le p$.*

Proof. For $\kappa < 0$, the assertion is clear. Suppose $0 \le \kappa \le p$. For an equivalence class $\widetilde{\mathfrak{F}} \in C^\kappa(\mathcal{T}_{\rm pr}(n+\ell,p+\ell),\varrho_{n+\ell,p+\ell})$ of fibers, there exists a proper Thom map $f : M \to N$ between manifolds of dimensions $n+\ell$ and $p+\ell$ such that its fiber over a point $y \in N$ is a representative of $\widetilde{\mathfrak{F}}$. By the proof of Lemma 7.3, we may assume that the stratum containing y is of codimension κ. Let B be a small open disk of dimension p embedded in N centered at y which is transverse to all the strata. Then $f|_{f^{-1}(B)} : f^{-1}(B) \to B$ is a proper Thom map and the ℓ-th suspension of its fiber over y is C^0 equivalent to the fiber of f over y by Thom's second isotopy lemma. Moreover, the codimension of the equivalence class containing the fiber of $f|_{f^{-1}(B)}$ over y is equal to κ. Hence, $s_\kappa(\widetilde{\mathfrak{F}})$ never vanishes. Since $\{\varrho_{p-k,p}\}_p$ is stable, this shows that s_κ is a monomorphism. □

Remark 8.7. We warn the reader that the equivalence class with respect to $\varrho_{n+\ell,p+\ell}$ of the ℓ-th suspension of a fiber whose equivalence class with respect to $\varrho_{n,p}$ is of codimension κ may not be of codimension κ. The codimension can decrease by suspension.

Remark 8.8. We see easily that for a κ with $0 \le \kappa \le p$, the $\mathbf{Z}_2$-linear map s_κ of (8.3) is an isomorphism if and only if the following two hold.

(1) If an equivalence class of fibers with respect to $\varrho_{n,p}$ has codimension κ, then the equivalence class of their ℓ-th suspensions with respect to $\varrho_{n+\ell,p+\ell}$ has also codimension κ.
(2) Two fibers whose equivalence classes with respect to $\varrho_{n,p}$ have codimension κ are equivalent with respect to $\varrho_{n,p}$ if and only if their ℓ-th suspensions are equivalent with respect to $\varrho_{n+\ell,p+\ell}$.

In particular, the $\mathbf{Z}_2$-linear maps s_κ are isomorphisms for all κ with $0 \leq \kappa \leq p$ if and only if the following holds: two fibers are equivalent with respect to $\varrho_{n,p}$ if and only if their ℓ-th suspensions are equivalent with respect to $\varrho_{n+\ell,p+\ell}$.

By virtue of Definition 8.5, we can prove the following.

Lemma 8.9. *The system of $\mathbf{Z}_2$-linear maps $\{s_\kappa\}_\kappa$ defines a cochain map*

$$\mathcal{C}(\mathcal{T}_{\mathrm{pr}}(n+\ell, p+\ell), \varrho_{n+\ell,p+\ell}) \to \mathcal{C}(\mathcal{T}_{\mathrm{pr}}(n,p), \varrho_{n,p}).$$

In other words, we have $\delta_\kappa \circ s_\kappa = s_{\kappa+1} \circ \delta_\kappa$ for all $\kappa \in \mathbf{Z}$.

Proof. We may assume that $0 \leq \kappa \leq p-1$. Let $\widetilde{\mathfrak{F}}$ be an equivalence class of fibers in $C^\kappa(\mathcal{T}_{\mathrm{pr}}(n+\ell, p+\ell), \varrho_{n+\ell,p+\ell})$, and $\widetilde{\mathfrak{G}}$ an equivalence class in $C^{\kappa+1}(\mathcal{T}_{\mathrm{pr}}(n,p), \varrho_{n,p})$. Let us consider the coefficients of $\widetilde{\mathfrak{G}}$ in $\delta_\kappa \circ s_\kappa(\widetilde{\mathfrak{F}})$ and in $s_{\kappa+1} \circ \delta_\kappa(\widetilde{\mathfrak{F}})$.

Case 1. The equivalence class of the ℓ-th suspension of $\widetilde{\mathfrak{G}}$ has codimension strictly smaller than $\kappa+1$.

The relevant coefficient in $s_{\kappa+1} \circ \delta_\kappa(\widetilde{\mathfrak{F}})$ is clearly zero by the definition of $s_{\kappa+1}$. On the other hand, if the relevant coefficient in $\delta_\kappa \circ s_\kappa(\widetilde{\mathfrak{F}})$ is not zero, then there is a codimension κ equivalence class $\widetilde{\mathfrak{H}}$ whose coboundary contains $\widetilde{\mathfrak{G}}$ and whose ℓ-th suspension is contained in $\widetilde{\mathfrak{F}}$. By our assumption, the ℓ-th suspension of $\widetilde{\mathfrak{G}}$ has codimension strictly smaller than $\kappa+1$, and hence either the ℓ-th suspension of $\widetilde{\mathfrak{H}}$ has codimension strictly smaller than κ, or the ℓ-th suspension of $\widetilde{\mathfrak{G}}$ is equivalent to the ℓ-th suspension of $\widetilde{\mathfrak{H}}$.

The first case does not occur, since the ℓ-th suspension of $\widetilde{\mathfrak{H}}$ is contained in $\widetilde{\mathfrak{F}}$, which is of codimension κ.

If the second case occurs, then the equivalence class of the ℓ-th suspension of $\widetilde{\mathfrak{G}}$ has codimension κ. Since by Lemma 7.3, the equivalence class determines a topological submanifold of codimension κ, there must be a unique codimension κ equivalence class $\widetilde{\mathfrak{H}}'$ $(\neq \widetilde{\mathfrak{H}})$ whose coboundary contains $\widetilde{\mathfrak{G}}$ and whose ℓ-th suspension is contained in $\widetilde{\mathfrak{F}}$. Hence, we see that the coefficient of $\widetilde{\mathfrak{G}}$ in $\delta_\kappa \circ s_\kappa(\widetilde{\mathfrak{F}})$ is equal to zero.

Hence, the relevant coefficients coincide with each other in this case.

Case 2. The equivalence class of the ℓ-th suspension of $\widetilde{\mathfrak{G}}$ has codimension $\kappa+1$.

The coefficient of $\widetilde{\mathfrak{G}}$ in $\delta_\kappa \circ s_\kappa(\widetilde{\mathfrak{F}})$ is equal to the number of codimension κ equivalence classes whose coboundaries contain $\widetilde{\mathfrak{G}}$ and whose ℓ-th suspensions are contained in $\widetilde{\mathfrak{F}}$. On the other hand, the coefficient of $\widetilde{\mathfrak{G}}$ in $s_{\kappa+1} \circ \delta_\kappa(\widetilde{\mathfrak{F}})$ is not zero if and only if the ℓ-th suspension of $\widetilde{\mathfrak{G}}$ is contained in the coboundary of $\widetilde{\mathfrak{F}}$. Hence, the relevant coefficients coincide with each other in this case as well. This completes the proof. □

It follows easily from the definition of s_κ that the composition of

$$s_\kappa : C^\kappa(\mathcal{T}_{\mathrm{pr}}(n+\ell+\ell', p+\ell+\ell'), \varrho_{n+\ell+\ell',p+\ell+\ell'}) \to C^\kappa(\mathcal{T}_{\mathrm{pr}}(n+\ell, p+\ell), \varrho_{n+\ell,p+\ell})$$

and

$$s_\kappa : C^\kappa(\mathcal{T}_{\mathrm{pr}}(n+\ell, p+\ell), \varrho_{n+\ell,p+\ell}) \to C^\kappa(\mathcal{T}_{\mathrm{pr}}(n, p), \varrho_{n,p})$$

coincides with

$$s_\kappa : C^\kappa(\mathcal{T}_{\mathrm{pr}}(n+\ell+\ell', p+\ell+\ell'), \varrho_{n+\ell+\ell',p+\ell+\ell'}) \to C^\kappa(\mathcal{T}_{\mathrm{pr}}(n, p), \varrho_{n,p}).$$

By this observation together with Lemma 8.9, for a fixed integer k, the projective limit

$$\mathcal{C}(\widetilde{\mathcal{T}}_{\mathrm{pr}}(k), \mathcal{R}_k) = \varprojlim_{p} \mathcal{C}(\mathcal{T}_{\mathrm{pr}}(p-k, p), \varrho_{p-k,p}) \tag{8.4}$$

is well-defined as a cochain complex. We call $\mathcal{C}(\widetilde{\mathcal{T}}_{\mathrm{pr}}(k), \mathcal{R}_k)$ the *universal complex of singular fibers for codimension k proper Thom maps with respect to the stable system of admissible equivalence relations $\mathcal{R}_k$*. We write its cohomology group of dimension κ by $H^\kappa(\widetilde{\mathcal{T}}_{\mathrm{pr}}(k), \mathcal{R}_k)$.

Remark 8.10. Recall that the projective limit (8.4) is identified with the subspace of the product

$$\prod_p C^\kappa(\mathcal{T}_{\mathrm{pr}}(p-k, p), \varrho_{p-k,p})$$

consisting of all elements $(c_p)_p$ with $s_\kappa(c_{p+\ell}) = c_p$ for all p and ℓ.

As a direct consequence of Lemmas 8.6 and 8.9, we have the following.

Lemma 8.11. *The natural map*

$$\Phi^\kappa_{n,p} : C^\kappa(\widetilde{\mathcal{T}}_{\mathrm{pr}}(k), \mathcal{R}_k) \to C^\kappa(\mathcal{T}_{\mathrm{pr}}(n, p), \varrho_{n,p}) \tag{8.5}$$

induced by the projection is a monomorphism if $\kappa \leq p$. Furthermore, the system of $\mathbf{Z}_2$-linear maps $\{\Phi^\kappa_{n,p}\}_\kappa$ defines a cochain map

$$\mathcal{C}(\widetilde{\mathcal{T}}_{\mathrm{pr}}(k), \mathcal{R}_k) \to \mathcal{C}(\mathcal{T}_{\mathrm{pr}}(n, p), \varrho_{n,p}).$$

The $\mathbf{Z}_2$-linear map $\Phi^\kappa_{n,p}$ defined above can be identified with the map (8.6) which will be defined in §8.4.

8.4 Another Description

The complex $\mathcal{C}(\widetilde{\mathcal{T}}_{\mathrm{pr}}(k), \mathcal{R}_k)$ can also be constructed by using another method, as explained below.

Definition 8.12. Let $f_i : M_i \to N_i$, $i = 0, 1$, be proper Thom maps with the same codimension $k = \dim N_i - \dim M_i$. We say that the fibers over $y_i \in N_i$, $i = 0, 1$, are *stably* C^0 (or C^∞) *equivalent* if the fibers of $f_i \times \mathrm{id}_{\mathbf{R}^{\ell_i}} : M_i \times \mathbf{R}^{\ell_i} \to N_i \times \mathbf{R}^{\ell_i}$ over $y_i \times \{0\}$ are C^0 (resp. C^∞) equivalent to each other for some nonnegative integers ℓ_i, $i = 0, 1$, with $\dim N_0 + \ell_0 = \dim N_1 + \ell_1$.

Definition 8.13. Suppose that an equivalence relation $\widehat{\mathcal{R}}_k$ among the fibers of proper Thom maps of codimension k is given. We say that the relation $\widehat{\mathcal{R}}_k$ is *stably admissible* if the following conditions are satisfied.

(1) If two fibers are stably C^0 equivalent, then they are also equivalent with respect to $\widehat{\mathcal{R}}_k$.
(2) For every positive integer ℓ, two fibers are equivalent with respect to $\widehat{\mathcal{R}}_k$ if and only if their ℓ-th suspensions are equivalent with respect to $\widehat{\mathcal{R}}_k$.
(3) For any proper Thom maps $f_i : M_i \to N_i$, $i = 0, 1$, of codimension k and for any points $y_i \in N_i$ whose corresponding fibers are equivalent with respect to $\widehat{\mathcal{R}}_k$, there exist neighborhoods U_i of $y_i \times \{0\}$ in $N_i \times \mathbf{R}^{\ell_i}$ for some nonnegative integers ℓ_i, $i = 0, 1$, with $\dim N_0 + \ell_0 = \dim N_1 + \ell_1$, and a homeomorphism $\varphi : U_0 \to U_1$ such that $\varphi(y_0 \times \{0\}) = y_1 \times \{0\}$ and
$$\varphi(U_0 \cap \widehat{\mathfrak{F}}(f_0 \times \mathrm{id}_{\mathbf{R}^{\ell_0}})) = U_1 \cap \widehat{\mathfrak{F}}(f_1 \times \mathrm{id}_{\mathbf{R}^{\ell_1}})$$
for every equivalence class $\widehat{\mathfrak{F}}$ of fibers with respect to $\widehat{\mathcal{R}}_k$, where $\widehat{\mathfrak{F}}(f_i \times \mathrm{id}_{\mathbf{R}^{\ell_i}})$ is the set of points in $N_i \times \mathbf{R}^{\ell_i}$ over which lies a fiber of $f_i \times \mathrm{id}_{\mathbf{R}^{\ell_i}}$ of type $\widehat{\mathfrak{F}}$.

For example, the stable C^0 equivalence is a stably admissible equivalence relation, and we denote it by $\widehat{\mathcal{R}}^0_k$.

The following lemma can be proved by an argument similar to that in the proof of Lemma 7.3.

Lemma 8.14. *For every equivalence class* $\widehat{\mathfrak{F}}$ *with respect to a stably admissible equivalence relation* $\widehat{\mathcal{R}}_k$, *and for every proper Thom map* $f : M \to N$ *in* $\widetilde{\mathcal{T}}_{\mathrm{pr}}(k)$, *the subspace* $\widehat{\mathfrak{F}}(f)$ *of* N *is a union of strata of* $\mathcal{N}$. *Furthermore, we have the following.*

(1) *For every* $y \in \widehat{\mathfrak{F}}(f)$, *there exists a nonnegative integer* ℓ *such that* $\widehat{\mathfrak{F}}(f) \times \mathbf{R}^\ell$ *is a* C^0 *submanifold of* $N \times \mathbf{R}^\ell$ *at* $y \times \{0\}$.
(2) *The codimension of* $\widehat{\mathfrak{F}}(f) \times \mathbf{R}^\ell$ *in* $N \times \mathbf{R}^\ell$ *at* $y \times \{0\}$ *does not depend on the choice of* y *or* f.

By virtue of the above lemma, the codimension of $\widehat{\mathfrak{F}}$ makes sense, and we denote it by $\kappa(\widehat{\mathfrak{F}})$.

Let $\widehat{\mathcal{R}}_k$ be a stably admissible equivalence relation among the fibers of proper Thom maps of codimension k. Then, we can naturally construct the cochain complex $\mathcal{C}(\widetilde{\mathcal{T}}_{\mathrm{pr}}(k), \widehat{\mathcal{R}}_k) = (C^\kappa(\widetilde{\mathcal{T}}_{\mathrm{pr}}(k), \widehat{\mathcal{R}}_k), \delta_\kappa)_\kappa$ as follows: the κ-dimensional cochain group $C^\kappa(\widetilde{\mathcal{T}}_{\mathrm{pr}}(k), \widehat{\mathcal{R}}_k)$ is the $\mathbf{Z}_2$-vector space consisting of all formal linear combinations, which may possibly contain infinitely many terms, of the equivalence classes $\widehat{\mathfrak{F}}$ of fibers of proper Thom maps of codimension k with $\kappa(\widehat{\mathfrak{F}}) = \kappa$ and

$$\delta_\kappa : C^\kappa(\widetilde{\mathcal{T}}_{\mathrm{pr}}(k), \widehat{\mathcal{R}}_k) \to C^{\kappa+1}(\widetilde{\mathcal{T}}_{\mathrm{pr}}(k), \widehat{\mathcal{R}}_k)$$

is defined in a way similar to $\delta_\kappa(f)$ (see (8.1) and the subsequent remark). (Here, we simply put $C^\kappa(\widetilde{\mathcal{T}}_{\mathrm{pr}}(k), \widehat{\mathcal{R}}_k) = 0$ for $\kappa < 0$.) Note that the incidence coefficient is well-defined by virtue of Definition 8.13 (2) and (3). We write the cohomology group of dimension κ of the cochain complex $\mathcal{C}(\widetilde{\mathcal{T}}_{\mathrm{pr}}(k), \widehat{\mathcal{R}}_k)$ by $H^\kappa(\widetilde{\mathcal{T}}_{\mathrm{pr}}(k), \widehat{\mathcal{R}}_k)$.

Let us now discuss the relationship between the complex thus obtained and that of §8.3. Suppose that a stable system $\mathcal{R}_k = \{\varrho_{p-k,p}\}_p$ of admissible equivalence relations for the fibers of proper Thom maps of codimension k in the sense of Definition 8.5 is given. Then, we can naturally define a new equivalence relation $\widehat{\mathcal{R}}_k$ for the fibers of proper Thom maps of codimension k as follows: two such fibers are equivalent if some of their suspensions are equivalent in the original sense. Then we can easily check that this new equivalence relation $\widehat{\mathcal{R}}_k$ is stably admissible in the sense of Definition 8.13. For example, if we consider the system of C^0 equivalence relations $\mathcal{R}_k^0 = \{\varrho_{p-k,p}^0\}_p$, then it defines a stable system of admissible equivalence relations, and the new equivalence relation is nothing but the stable C^0 equivalence $\widehat{\mathcal{R}}_k^0$.

Then, we get the following.

Proposition 8.15. *The complex $\mathcal{C}(\widetilde{\mathcal{T}}_{\mathrm{pr}}(k), \widehat{\mathcal{R}}_k)$ with respect to the new equivalence relation $\widehat{\mathcal{R}}_k$ defined above is naturally isomorphic to the universal complex $\mathcal{C}(\widetilde{\mathcal{T}}_{\mathrm{pr}}(k), \mathcal{R}_k)$, defined by (8.4), of singular fibers for codimension k proper Thom maps with respect to the original stable system of admissible equivalence relations $\mathcal{R}_k = \{\varrho_{p-k,p}\}_p$.*

Proof. For every pair (n, p) with $p - n = k$ and for every κ, we can naturally define the $\mathbf{Z}_2$-linear map

$$\Phi_{n,p}^\kappa : C^\kappa(\widetilde{\mathcal{T}}_{\mathrm{pr}}(k), \widehat{\mathcal{R}}_k) \to C^\kappa(\mathcal{T}_{\mathrm{pr}}(n,p), \varrho_{n,p}) \tag{8.6}$$

by associating to each equivalence class $\widehat{\mathfrak{F}}$ of codimension κ with respect to $\widehat{\mathcal{R}}_k$ the sum of all those equivalence classes of codimension κ with respect to $\varrho_{n,p}$ which are contained in $\widehat{\mathfrak{F}}$. It is not difficult to show that $\Phi_{n,p} = \{\Phi_{n,p}^\kappa\}_\kappa$ defines a cochain map

$$\mathcal{C}(\widetilde{\mathcal{T}}_{\mathrm{pr}}(k), \widehat{\mathcal{R}}_k) \to \mathcal{C}(\mathcal{T}_{\mathrm{pr}}(n,p), \varrho_{n,p})$$

(see the proof of Lemma 8.9) and that $s_\kappa \circ \Phi_{n+\ell,p+\ell}^\kappa = \Phi_{n,p}^\kappa$ for every positive integer ℓ, where s_κ is the $\mathbf{Z}_2$-linear map (8.3) induced by the suspension. Hence, $\{\Phi_{p-k,p}\}_p$ induces a cochain map

$$\Phi : \mathcal{C}(\widetilde{\mathcal{T}}_{\mathrm{pr}}(k), \widehat{\mathcal{R}}_k) \to \varprojlim_p \mathcal{C}(\mathcal{T}_{\mathrm{pr}}(p-k,p), \varrho_{p-k,p}) = \mathcal{C}(\widetilde{\mathcal{T}}_{\mathrm{pr}}(k), \mathcal{R}_k)$$

by the universality of the projective limit. Furthermore, it is not difficult to show that Φ is injective. Finally, Φ is surjective by virtue of the definitions of $\widehat{\mathcal{R}}_k$ and the projective limit. Hence, we have the desired conclusion. This completes the proof. □

Conversely, suppose that a stably admissible equivalence relation $\widehat{\mathcal{R}}_k$ among the fibers of proper Thom maps of codimension k is given. Then, for every pair (n, p) with $p - n = k$, we can define the equivalence relation $\varrho_{n,p}$ among the fibers of elements of $\mathcal{T}_{\mathrm{pr}}(n, p)$ as follows: two such fibers are equivalent with respect to $\varrho_{n,p}$ if they are equivalent with respect to $\widehat{\mathcal{R}}_k$ and in Definition 8.13 (3), ℓ_i can be chosen to be zero, i.e., if there exist neighborhoods U_i of y_i in N_i, $i = 0, 1$, and a homeomorphism $\varphi : U_0 \to U_1$ such that $\varphi(y_0) = y_1$ and

$$\varphi(U_0 \cap \widehat{\mathfrak{F}}(f_0)) = U_1 \cap \widehat{\mathfrak{F}}(f_1)$$

for every equivalence class $\widehat{\mathfrak{F}}$ of fibers with respect to $\widehat{\mathcal{R}}_k$, where $f_i : M_i \to N_i$ are elements of $\mathcal{T}_{\mathrm{pr}}(n, p)$ whose fibers over $y_i \in N_i$ are the given ones, and $\widehat{\mathfrak{F}}(f_i)$ is the set of points in N_i over which lies a fiber of f_i of type $\widehat{\mathfrak{F}}$.

Lemma 8.16. (1) *The relation $\varrho_{n,p}$ defined as above is an admissible equivalence relation in the sense of Definition* 7.2.

(2) *The system of equivalence relations $\mathcal{R}_k = \{\varrho_{p-k,p}\}_p$ is stable in the sense of Definition* 8.5.

Proof. (1) We can show that the C^0 equivalence implies the equivalence with respect to $\varrho_{n,p}$, since C^0 equivalence implies the equivalence with respect to $\widehat{\mathcal{R}}_k$.

Suppose that $f_i : M_i \to N_i$, $i = 0, 1$, are elements of $\mathcal{T}_{\mathrm{pr}}(n, p)$ whose fibers over $y_i \in N_i$ are equivalent to each other with respect to $\varrho_{n,p}$. Then, there exist neighborhoods U_i of y_i in N_i, $i = 0, 1$, and a homeomorphism $\varphi : U_0 \to U_1$ such that $\varphi(y_0) = y_1$ and $\varphi(U_0 \cap \widehat{\mathfrak{F}}(f_0)) = U_1 \cap \widehat{\mathfrak{F}}(f_1)$ for every equivalence class $\widehat{\mathfrak{F}}$ of fibers with respect to $\widehat{\mathcal{R}}_k$. Then, the fiber of f_0 over an arbitrary point $y \in U_0$ is equivalent, with respect to $\varrho_{n,p}$, to that of f_1 over $\varphi(y)$, by the very definition of the equivalence relation $\varrho_{n,p}$. Hence, we have $\varphi(U_0 \cap \widetilde{\mathfrak{F}}(f_0)) = U_1 \cap \widetilde{\mathfrak{F}}(f_1)$ for every equivalence class $\widetilde{\mathfrak{F}}$ with respect to $\varrho_{n,p}$. Hence, (1) holds.

(2) This follows immediately from the definition of the equivalence relation $\varrho_{n,p}$. This completes the proof. □

We see easily that the stably admissible equivalence relation among the fibers of proper Thom maps of codimension k constructed from $\mathcal{R}_k$ coincides with the original equivalence relation $\widehat{\mathcal{R}}_k$. Therefore, by Proposition 8.15, the complex $\mathcal{C}(\widetilde{\mathcal{T}}_{\mathrm{pr}}(k), \widehat{\mathcal{R}}_k)$ is naturally isomorphic to the universal complex $\mathcal{C}(\widetilde{\mathcal{T}}_{\mathrm{pr}}(k), \mathcal{R}_k)$, defined by (8.4), of singular fibers for codimension k proper Thom maps with respect to the stable system of admissible equivalence relations $\mathcal{R}_k = \{\varrho_{p-k,p}\}_p$.

For the stable C^0 equivalence, we have the following problem, the answer to which the author does not know.

Problem 8.17. Let $f_i : M_i \to N_i$, $i = 0, 1$, be proper Thom maps such that $n = \dim M_0 = \dim M_1$ and $p = \dim N_0 = \dim N_1$. For points $y_i \in N_i$, if the fibers of f_i over y_i are stably C^0 (or C^∞) equivalent, then are they C^0 (resp. C^∞) equivalent? In other words, is the natural cochain map

$$\mathcal{C}(\widetilde{\mathcal{T}}_{\mathrm{pr}}(k), \mathcal{R}_k^0) \to \mathcal{C}(\mathcal{T}_{\mathrm{pr}}(n,p), \varrho_{n,p}^0)$$

of the universal complex with respect to the stable C^0 equivalence to that with respect to the C^0 equivalence an epimorphism?

Note that if f_i are codimension -1 proper C^0 stable maps of manifolds of dimension less than or equal to 4, then the answer to the above problem is shown to be affirmative by using an argument similar to that in the proof of Corollary 3.9. (In the 4-dimensional case, we should assume the orientability of the source manifold, while for the other dimensions, it is not necessary. See Corollary 3.16 and the subsequent remark.)

8.5 Changing the Equivalence Relation

Suppose that we are given two admissible equivalence relations $\varrho = \varrho_{n,p}$ and $\overline{\varrho} = \overline{\varrho}_{n,p}$ for the fibers of elements of $\mathcal{T}_{\mathrm{pr}}(n,p)$. If every equivalence class with respect to $\varrho_{n,p}$ is a union of equivalence classes with respect to $\overline{\varrho}_{n,p}$, then we say that $\varrho_{n,p}$ is *weaker* than $\overline{\varrho}_{n,p}$ and write $\varrho_{n,p} \leq \overline{\varrho}_{n,p}$. In this case, we can naturally define the $\mathbf{Z}_2$-linear map

$$\varepsilon_{\varrho,\overline{\varrho}} : \mathcal{C}(\mathcal{T}_{\mathrm{pr}}(n,p), \varrho) \to \mathcal{C}(\mathcal{T}_{\mathrm{pr}}(n,p), \overline{\varrho}) \tag{8.7}$$

by associating to a class $\widetilde{\mathfrak{F}}$ of codimension κ with respect to ϱ the sum of all the equivalence classes with respect to $\overline{\varrho}$ of codimension κ contained in $\widetilde{\mathfrak{F}}$. This clearly defines a cochain map (for example, see the proof of Lemma 8.9). Note that the associated map

$$C^\kappa(\mathcal{T}_{\mathrm{pr}}(n,p), \varrho) \to C^\kappa(\mathcal{T}_{\mathrm{pr}}(n,p), \overline{\varrho})$$

is a monomorphism for every κ.

Suppose that we are given two stable systems of admissible equivalence relations $\mathcal{R}_k = \{\varrho_{p-k,p}\}_p$ and $\overline{\mathcal{R}}_k = \{\overline{\varrho}_{p-k,p}\}_p$ for the fibers of codimension k proper Thom maps. If $\varrho_{p-k,p} \leq \overline{\varrho}_{p-k,p}$ for every p, then we say that $\mathcal{R}_k$ is *weaker* than $\overline{\mathcal{R}}_k$ and write $\mathcal{R}_k \leq \overline{\mathcal{R}}_k$. In this case, the system of cochain maps $\{\varepsilon_{\varrho_{p-k,p},\overline{\varrho}_{p-k,p}}\}_p$ induces the cochain map

$$\varepsilon_{\mathcal{R}_k,\overline{\mathcal{R}}_k} : \mathcal{C}(\widetilde{\mathcal{T}}_{\mathrm{pr}}(k), \mathcal{R}_k) \to \mathcal{C}(\widetilde{\mathcal{T}}_{\mathrm{pr}}(k), \overline{\mathcal{R}}_k).$$

Note that the associated map

$$C^{\kappa}(\widetilde{\mathcal{T}}_{\mathrm{pr}}(k), \mathcal{R}_k) \to C^{\kappa}(\widetilde{\mathcal{T}}_{\mathrm{pr}}(k), \overline{\mathcal{R}}_k)$$

is a monomorphism for every κ.

In particular, if we consider the C^0 equivalence $\varrho^0_{n,p}$ among the fibers of elements of $\mathcal{T}_{\mathrm{pr}}(n,p)$, then we have $\varrho_{n,p} \leq \varrho^0_{n,p}$ for any admissible equivalence relation $\varrho_{n,p}$. Hence, we have the cochain map

$$\varepsilon_{\varrho_{n,p},\varrho^0_{n,p}} : \mathcal{C}(\mathcal{T}_{\mathrm{pr}}(n,p), \varrho_{n,p}) \to \mathcal{C}(\mathcal{T}_{\mathrm{pr}}(n,p), \varrho^0_{n,p}).$$

In other words, since this cochain map is always a monomorphism, we may regard $\mathcal{C}(\mathcal{T}_{\mathrm{pr}}(n,p), \varrho_{n,p})$ as a subcomplex of $\mathcal{C}(\mathcal{T}_{\mathrm{pr}}(n,p), \varrho^0_{n,p})$.

Furthermore, if we consider the stable system of admissible equivalence relations $\mathcal{R}^0_k = \{\varrho^0_{p-k,p}\}_p$ induced by the C^0 equivalence, then $\mathcal{R}_k \leq \mathcal{R}^0_k$ for any stable system $\mathcal{R}_k$ of admissible equivalence relations. Hence, we have the cochain map

$$\varepsilon_{\mathcal{R}_k,\mathcal{R}^0_k} : \mathcal{C}(\widetilde{\mathcal{T}}_{\mathrm{pr}}(k), \mathcal{R}_k) \to \mathcal{C}(\widetilde{\mathcal{T}}_{\mathrm{pr}}(k), \mathcal{R}^0_k).$$

We can show that this is always a monomorphism, and hence $\mathcal{C}(\widetilde{\mathcal{T}}_{\mathrm{pr}}(k), \mathcal{R}_k)$ can be regarded as a subcomplex of $\mathcal{C}(\widetilde{\mathcal{T}}_{\mathrm{pr}}(k), \mathcal{R}^0_k)$.

8.6 Changing the Class of Maps

So far, we have worked with the whole set of proper Thom maps of a fixed codimension. By restricting the class of Thom maps that we consider, we can also obtain the universal complex of singular fibers for such a class of maps.

First, let us consider maps between manifolds of fixed dimensions.

Definition 8.18. A C^0 equivalence class $\mathfrak{F}$ of fibers of elements of $\mathcal{T}_{\mathrm{pr}}(n,p)$ is said to be *under* another C^0 equivalence class $\mathfrak{G}$ of fibers if for some (and hence every) representative $f : (M, f^{-1}(y)) \to (N, y)$ of $\mathfrak{F}$, there is a point y' arbitrarily close to y over which lies a fiber of type $\mathfrak{G}$. In this case, we also say that $\mathfrak{G}$ is *over* $\mathfrak{F}$.

Let $\Gamma = \Gamma_{n,p}$ be an ascending set of C^0 equivalence classes of fibers of elements of $\mathcal{T}_{\mathrm{pr}}(n,p)$, where "*ascending*" means that for an arbitrary equivalence class in the set, every class over it also belongs to the set. We say that a proper Thom map $f : M \to N$ between smooth manifolds of dimensions n and p is a Γ*-map* if its fibers all lie in Γ. We use the same notation $\Gamma = \Gamma_{n,p}$ for the set of all Γ-maps, when there is no confusion.

If for an arbitrary equivalence class in the set Γ, every class under it also belongs to the set, then we say that it is a *descending* set. For example, the set of all C^0 equivalence classes of fibers of a fixed Thom map $f \in \mathcal{T}_{\mathrm{pr}}(n,p)$ is ascending, while the set f^c of all C^0 equivalence classes of fibers of elements of $\mathcal{T}_{\mathrm{pr}}(n,p)$ which do not appear for f is a descending set. Note that $\mathcal{C}(f^c, \varrho^0)$ is a subcomplex of $\mathcal{C}(\mathcal{T}_{\mathrm{pr}}(n,p), \varrho^0)$ (see Lemma 8.2), essentially because the set f^c is descending.

Let $\Gamma = \Gamma_{n,p}$ be as above and let $\varrho^{\Gamma} = \varrho^{\Gamma}_{n,p}$ be an equivalence relation among the fibers of Γ-maps which is admissible in the same sense as in Definition 7.2. Then, we can naturally define the universal complex $\mathcal{C}(\Gamma_{n,p}, \varrho^{\Gamma})$ of singular fibers for Γ-maps with respect to the admissible equivalence relation ϱ^{Γ}. We write the corresponding cohomology group of dimension κ by $H^{\kappa}(\Gamma_{n,p}, \varrho^{\Gamma})$.

Suppose that the equivalence relation ϱ^{Γ} is the restriction to Γ of an admissible equivalence relation $\varrho = \varrho_{n,p}$ among the fibers of elements of $\mathcal{T}_{\mathrm{pr}}(n,p)$. Let $C^{\kappa}(\Gamma^c, \varrho)$ be the linear subspace of $C^{\kappa}(\mathcal{T}_{\mathrm{pr}}(n,p), \varrho)$ spanned by those equivalence classes of fibers of elements of $\mathcal{T}_{\mathrm{pr}}(n,p)$ of codimension κ with respect to ϱ which contain no fiber of a Γ-map. Then, by an argument similar to that in the proof of Lemma 8.2, we can prove the following. Details are left to the reader.

Lemma 8.19. *For an ascending set $\Gamma = \Gamma_{n,p}$ of C^0 equivalence classes of fibers of elements of $\mathcal{T}_{\mathrm{pr}}(n,p)$, the following holds.*

(1) *We have $\delta_{\kappa}(C^{\kappa}(\Gamma^c, \varrho)) \subset C^{\kappa+1}(\Gamma^c, \varrho)$ for every $\kappa \in \mathbf{Z}$. Hence, $\mathcal{C}(\Gamma^c, \varrho) = (C^{\kappa}(\Gamma^c, \varrho), \delta_{\kappa}|_{C^{\kappa}(\Gamma^c,\varrho)})_{\kappa}$ constitutes a subcomplex of $\mathcal{C}(\mathcal{T}_{\mathrm{pr}}(n,p), \varrho)$.*
(2) *The quotient complex*

$$\mathcal{C}(\mathcal{T}_{\mathrm{pr}}(n,p), \varrho)/\mathcal{C}(\Gamma^c, \varrho) = (C^{\kappa}(\mathcal{T}_{\mathrm{pr}}(n,p), \varrho)/C^{\kappa}(\Gamma^c, \varrho), \overline{\delta}_{\kappa})_{\kappa}$$

is naturally isomorphic to $\mathcal{C}(\Gamma, \varrho^{\Gamma})$, where

$$\overline{\delta}_{\kappa} : C^{\kappa}(\mathcal{T}_{\mathrm{pr}}(n,p), \varrho)/C^{\kappa}(\Gamma^c, \varrho) \to C^{\kappa+1}(\mathcal{T}_{\mathrm{pr}}(n,p), \varrho)/C^{\kappa+1}(\Gamma^c, \varrho)$$

is the well-defined $\mathbf{Z}_2$-linear map induced by δ_{κ}.

More generally, if Γ and Γ' are two ascending sets of singular fibers of elements of $\mathcal{T}_{\mathrm{pr}}(n,p)$ such that $\Gamma \subset \Gamma'$, and if the admissible equivalence relation ϱ^{Γ} for Γ is the restriction to Γ of an admissible equivalence relation $\varrho^{\Gamma'}$ for Γ', then we naturally have the cochain map

$$\pi_{\Gamma',\Gamma} : \mathcal{C}(\Gamma', \varrho^{\Gamma'}) \to \mathcal{C}(\Gamma, \varrho^{\Gamma})$$

induced by the projection. Note that the corresponding $\mathbf{Z}_2$-linear map on every dimension is an epimorphism.

Furthermore, if $\varrho^{\Gamma'}$ and $\overline{\varrho}^{\Gamma'}$ are two admissible equivalence relations for Γ' with $\varrho^{\Gamma'} \leq \overline{\varrho}^{\Gamma'}$ in a sense similar to §8.5, then we can naturally define the cochain map

$$\varepsilon_{\varrho^{\Gamma'},\overline{\varrho}^{\Gamma'}} : \mathcal{C}(\Gamma', \varrho^{\Gamma'}) \to \mathcal{C}(\Gamma', \overline{\varrho}^{\Gamma'}).$$

Note that the corresponding $\mathbf{Z}_2$-linear map on every dimension is a monomorphism. If ϱ^{Γ} and $\overline{\varrho}^{\Gamma}$ are the restrictions to Γ of $\varrho^{\Gamma'}$ and $\overline{\varrho}^{\Gamma'}$ respectively, then we have the commutative diagram of cochain complexes:

$$\begin{array}{ccc} \mathcal{C}(\Gamma', \varrho^{\Gamma'}) & \xrightarrow{\varepsilon_{\varrho^{\Gamma'}, \overline{\varrho}^{\Gamma'}}} & \mathcal{C}(\Gamma', \overline{\varrho}^{\Gamma'}) \\ {\scriptstyle \pi_{\Gamma',\Gamma}} \downarrow & & \downarrow {\scriptstyle \pi_{\Gamma',\Gamma}} \\ \mathcal{C}(\Gamma, \varrho^{\Gamma}) & \xrightarrow{\varepsilon_{\varrho^{\Gamma}, \overline{\varrho}^{\Gamma}}} & \mathcal{C}(\Gamma, \overline{\varrho}^{\Gamma}). \end{array} \tag{8.8}$$

Let us denote by $\mathcal{C}(\Gamma' \setminus \Gamma, \varrho^{\Gamma'})$ and $\mathcal{C}(\Gamma' \setminus \Gamma, \overline{\varrho}^{\Gamma'})$ the kernels of the $\mathbf{Z}_2$-linear maps

$$\pi_{\Gamma',\Gamma} : \mathcal{C}(\Gamma', \varrho^{\Gamma'}) \to \mathcal{C}(\Gamma, \varrho^{\Gamma})$$

and

$$\pi_{\Gamma',\Gamma} : \mathcal{C}(\Gamma', \overline{\varrho}^{\Gamma'}) \to \mathcal{C}(\Gamma, \overline{\varrho}^{\Gamma})$$

respectively. Note that they are subcomplexes of $\mathcal{C}(\Gamma', \varrho^{\Gamma'})$ and $\mathcal{C}(\Gamma', \overline{\varrho}^{\Gamma'})$ respectively spanned by those equivalence classes of fibers in Γ' which contain no fiber in Γ. Furthermore, we define $\mathcal{C}(\Gamma, \overline{\varrho}^{\Gamma}/\varrho^{\Gamma})$ and $\mathcal{C}(\Gamma', \overline{\varrho}^{\Gamma'}/\varrho^{\Gamma'})$ to be the cokernels of $\varepsilon_{\varrho^{\Gamma}, \overline{\varrho}^{\Gamma}}$ and $\varepsilon_{\varrho^{\Gamma'}, \overline{\varrho}^{\Gamma'}}$ respectively. It is easy to show that $\varepsilon_{\varrho^{\Gamma'}, \overline{\varrho}^{\Gamma'}}$ induces a monomorphism

$$\mathcal{C}(\Gamma' \setminus \Gamma, \varrho^{\Gamma'}) \to \mathcal{C}(\Gamma' \setminus \Gamma, \overline{\varrho}^{\Gamma'})$$

and we denote its cokernel by $\mathcal{C}(\Gamma' \setminus \Gamma, \overline{\varrho}^{\Gamma'}/\varrho^{\Gamma'})$. Then we naturally have the following commutative diagram:

$$\begin{array}{ccccccccc} & & 0 & & 0 & & 0 & & \\ & & \downarrow & & \downarrow & & \downarrow & & \\ 0 & \to & \mathcal{C}(\Gamma' \setminus \Gamma, \varrho^{\Gamma'}) & \to & \mathcal{C}(\Gamma' \setminus \Gamma, \overline{\varrho}^{\Gamma'}) & \to & \mathcal{C}(\Gamma' \setminus \Gamma, \overline{\varrho}^{\Gamma'}/\varrho^{\Gamma'}) & \to & 0 \\ & & \downarrow & & \downarrow & & \downarrow & & \\ 0 & \to & \mathcal{C}(\Gamma', \varrho^{\Gamma'}) & \xrightarrow{\varepsilon_{\varrho^{\Gamma'}, \overline{\varrho}^{\Gamma'}}} & \mathcal{C}(\Gamma', \overline{\varrho}^{\Gamma'}) & \to & \mathcal{C}(\Gamma', \overline{\varrho}^{\Gamma'}/\varrho^{\Gamma'}) & \to & 0 \\ & & {\scriptstyle \pi_{\Gamma',\Gamma}} \downarrow & & \downarrow {\scriptstyle \pi_{\Gamma',\Gamma}} & & \downarrow & & \\ 0 & \to & \mathcal{C}(\Gamma, \varrho^{\Gamma}) & \xrightarrow{\varepsilon_{\varrho^{\Gamma}, \overline{\varrho}^{\Gamma}}} & \mathcal{C}(\Gamma, \overline{\varrho}^{\Gamma}) & \to & \mathcal{C}(\Gamma, \overline{\varrho}^{\Gamma}/\varrho^{\Gamma}) & \to & 0 \\ & & \downarrow & & \downarrow & & \downarrow & & \\ & & 0 & & 0 & & 0 & & \end{array} \tag{8.9}$$

where all the rows and columns are exact. Therefore, we have the corresponding commutative diagram of long exact sequences of cohomologies as well.

Now let us vary the dimension pair (n, p) keeping the codimension $p-n = k$ fixed. Let

$$\widetilde{\Gamma} = \widetilde{\Gamma}_k = \bigcup_{p-n=k} \Gamma_{n,p}$$

be a set of C^0 equivalence classes of fibers of proper Thom maps of codimension k such that each $\Gamma_{n,p}$ is an ascending set of C^0 equivalence classes of fibers of elements of $\mathcal{T}_{\mathrm{pr}}(n, p)$, and that $\widetilde{\Gamma}$ is closed under suspension in the sense of Definition 8.4. (For example, the set of all C^0 equivalence classes of fibers of elements of $\widetilde{\mathcal{S}}^0_{\mathrm{pr}}(k)$ is such a set.)

We say that a proper Thom map of codimension k is a $\widetilde{\Gamma}_k$*-map* if its fibers all lie in $\widetilde{\Gamma}_k$. We use the same notation $\widetilde{\Gamma} = \widetilde{\Gamma}_k$ for the set of all $\widetilde{\Gamma}_k$-maps, when there is no confusion.

Let $\mathcal{R}_k^{\widetilde{\Gamma}} = \{\varrho_{p-k,p}^{\Gamma_{p-k,p}}\}_p$ be a system of equivalence relations, where each $\varrho_{p-k,p}^{\Gamma_{p-k,p}}$ is an admissible equivalence relation among the fibers of $\Gamma_{p-k,p}$-maps. Furthermore, we assume that the system $\mathcal{R}_k^{\widetilde{\Gamma}}$ of admissible equivalence relations is stable in the sense of Definition 8.5.

Then, we can naturally define the universal complex of singular fibers

$$\mathcal{C}(\widetilde{\Gamma}_k, \mathcal{R}_k^{\widetilde{\Gamma}})$$

for $\widetilde{\Gamma}_k$-maps with respect to the stable system of admissible equivalence relations $\mathcal{R}_k^{\widetilde{\Gamma}}$. As in the case of Thom maps, we have two definitions for the universal complexes, which are equivalent to each other as in Proposition 8.15. We write its cohomology group of dimension κ by $H^\kappa(\widetilde{\Gamma}_k, \mathcal{R}_k^{\widetilde{\Gamma}})$.

Note that if the stable system of admissible equivalence relations $\mathcal{R}_k^{\widetilde{\Gamma}}$ is the restriction of a stable system of admissible equivalence relations $\mathcal{R}_k$ among the fibers of proper Thom maps of codimension k, then we see that the complex $\mathcal{C}(\widetilde{\Gamma}_k, \mathcal{R}_k^{\widetilde{\Gamma}})$ is a quotient complex of the universal complex $\mathcal{C}(\widetilde{\mathcal{T}}_{\mathrm{pr}}(k), \mathcal{R}_k)$ in view of the construction given in §8.4.

More generally, if $\widetilde{\Gamma}$ and $\widetilde{\Gamma}'$ are two ascending sets of singular fibers of elements of $\widetilde{\mathcal{T}}_{\mathrm{pr}}(k)$ which are closed under suspension such that $\widetilde{\Gamma} \subset \widetilde{\Gamma}'$, and if the stable system of admissible equivalence relations $\mathcal{R}_k^{\widetilde{\Gamma}}$ for $\widetilde{\Gamma}$ is the restriction to $\widetilde{\Gamma}$ of a stable system of admissible equivalence relations $\mathcal{R}_k^{\widetilde{\Gamma}'}$ for $\widetilde{\Gamma}'$, then we naturally have the cochain map

$$\pi_{\widetilde{\Gamma}', \widetilde{\Gamma}} : \mathcal{C}(\widetilde{\Gamma}', \mathcal{R}_k^{\widetilde{\Gamma}'}) \to \mathcal{C}(\widetilde{\Gamma}, \mathcal{R}_k^{\widetilde{\Gamma}})$$

induced by the natural projection. Note that the corresponding $\mathbf{Z}_2$-linear map on every dimension is an epimorphism.

Furthermore, if $\mathcal{R}^{\widetilde{\Gamma}'}$ and $\overline{\mathcal{R}}^{\widetilde{\Gamma}'}$ are two stable systems of admissible equivalence relations for $\widetilde{\Gamma}'$ with $\mathcal{R}^{\widetilde{\Gamma}'} \leq \overline{\mathcal{R}}^{\widetilde{\Gamma}'}$ in a sense similar to §8.5, then we can naturally define the cochain map

$$\varepsilon_{\mathcal{R}^{\widetilde{\Gamma}'}, \overline{\mathcal{R}}^{\widetilde{\Gamma}'}} : \mathcal{C}(\widetilde{\Gamma}', \mathcal{R}^{\widetilde{\Gamma}'}) \to \mathcal{C}(\widetilde{\Gamma}', \overline{\mathcal{R}}^{\widetilde{\Gamma}'}).$$

Note that the corresponding $\mathbf{Z}_2$-linear map on every dimension is a monomorphism.

If $\mathcal{R}^{\widetilde{\Gamma}}$ and $\overline{\mathcal{R}}^{\widetilde{\Gamma}}$ are the restrictions to $\widetilde{\Gamma}$ of $\mathcal{R}^{\widetilde{\Gamma}'}$ and $\overline{\mathcal{R}}^{\widetilde{\Gamma}'}$ respectively, then we have the commutative diagram of cochain maps:

$$\begin{array}{ccc}
\mathcal{C}(\widetilde{\Gamma}', \mathcal{R}^{\widetilde{\Gamma}'}) & \xrightarrow{\varepsilon_{\mathcal{R}^{\widetilde{\Gamma}'}, \overline{\mathcal{R}}^{\widetilde{\Gamma}'}}} & \mathcal{C}(\widetilde{\Gamma}', \overline{\mathcal{R}}^{\widetilde{\Gamma}'}) \\
{\scriptstyle \pi_{\widetilde{\Gamma}', \widetilde{\Gamma}}} \downarrow & & \downarrow {\scriptstyle \pi_{\widetilde{\Gamma}', \widetilde{\Gamma}}} \\
\mathcal{C}(\widetilde{\Gamma}, \mathcal{R}^{\widetilde{\Gamma}}) & \xrightarrow{\varepsilon_{\mathcal{R}^{\widetilde{\Gamma}}, \overline{\mathcal{R}}^{\widetilde{\Gamma}}}} & \mathcal{C}(\widetilde{\Gamma}, \overline{\mathcal{R}}^{\widetilde{\Gamma}}).
\end{array}$$

Note that we can extend the above commutative diagram as in (8.9) so that we obtain exact rows and columns.

Remark 8.20. Let $\widetilde{\Gamma} = \widetilde{\Gamma}_k = \cup_p \Gamma_{p-k,p}$ be as above and $\mathcal{R}_k^{\widetilde{\Gamma}} = \{\varrho_{p-k,p}^{\Gamma_{p-k,p}}\}_p$ be a stable system of admissible equivalence relations for the fibers of $\widetilde{\Gamma}$-maps. Then we have the natural $\mathbf{Z}_2$-linear map

$$\Phi_{p-k,p}^{\kappa} : C^{\kappa}(\widetilde{\Gamma}, \mathcal{R}_k^{\widetilde{\Gamma}}) \to C^{\kappa}(\Gamma_{p-k,p}, \varrho_{p-k,p}^{\Gamma_{p-k,p}})$$

induced by the projection for every p, since $C^{\kappa}(\widetilde{\Gamma}, \mathcal{R}_k^{\widetilde{\Gamma}})$ is the projective limit and hence is a $\mathbf{Z}_2$-submodule of the product of all $C^{\kappa}(\Gamma_{p-k,p}, \varrho_{p-k,p}^{\Gamma_{p-k,p}})$ (see Remark 8.10). (Note that for $\widetilde{\Gamma} = \widetilde{\mathcal{T}}_{\mathrm{pr}}(k)$, this map has already been defined. See (8.5) and (8.6).) Set $n = p - k$. For $0 \leq \kappa \leq p$, $\Phi_{n,p}^{\kappa}$ is a monomorphism if and only if every equivalence class of fibers in $\widetilde{\Gamma}$ with respect to $\widehat{\mathcal{R}}_k$ of codimension κ contains a suspension of a fiber in $\Gamma_{p-k,p}$ whose equivalence class with respect to $\varrho_{p-k,p}^{\Gamma_{p-k,p}}$ has codimension κ, where $\widehat{\mathcal{R}}_k$ is the stably admissible equivalence relation among the fibers in $\widetilde{\Gamma}$ defined just before Proposition 8.15 (compare this assertion with Lemma 8.11). On the other hand, $\Phi_{n,p}^{\kappa}$ is an epimorphism if and only if the following two conditions hold.

(1) If an equivalence class of fibers in $\Gamma_{n,p}$ with respect to $\varrho_{n,p}^{\Gamma_{n,p}}$ has codimension κ, then the equivalence class of their ℓ-th suspensions with respect to $\varrho_{n+\ell,p+\ell}^{\Gamma_{n+\ell,p+\ell}}$ has also codimension κ for all $\ell \geq 1$.
(2) Two fibers in $\Gamma_{n,p}$ whose equivalence classes with respect to $\varrho_{n,p}^{\Gamma_{n,p}}$ have codimension κ are equivalent with respect to $\varrho_{n,p}^{\Gamma_{n,p}}$ if and only if their ℓ-th suspensions are equivalent with respect to $\varrho_{n+\ell,p+\ell}^{\Gamma_{n+\ell,p+\ell}}$ for some $\ell \geq 0$.

Compare this with Problem 8.17, Lemma 8.6 and Remark 8.8.

When a class of proper Thom maps is given, let us consider the following definitions.

Definition 8.21. (1) Let $\Gamma_{n,p} = \Gamma$ be a subset of $\mathcal{T}_{\mathrm{pr}}(n,p)$. We denote by $\Gamma_{n,p}^* = \Gamma^*$ the set of all C^0 equivalence classes of fibers of elements of $\Gamma_{n,p}$. Then, it is clear that $\Gamma_{n,p}^*$ is an ascending set and the set of all $\Gamma_{n,p}^*$-maps contain the original set $\Gamma_{n,p}$ of maps. For an admissible equivalence relation ϱ^{Γ} among the elements of $\Gamma_{n,p}^*$, we define the *universal complex of singular fibers for* $\Gamma_{n,p}$ *with respect to* ϱ^{Γ} by

$$\mathcal{C}(\Gamma_{n,p}, \varrho^{\Gamma}) = \mathcal{C}(\Gamma_{n,p}^*, \varrho^{\Gamma}).$$

Furthermore, we denote the corresponding cohomology group of dimension κ by $H^{\kappa}(\Gamma_{n,p}, \varrho^{\Gamma})$. We call the set of $\Gamma_{n,p}^*$-maps the *completion* of $\Gamma_{n,p}$. When the set of $\Gamma_{n,p}^*$-maps coincides with the original set $\Gamma_{n,p}$, we say that the set $\Gamma_{n,p}$ is *complete*.

(2) Let $\widetilde{\Gamma}_k = \widetilde{\Gamma}$ be a subset of $\widetilde{\mathcal{T}}_{\rm pr}(k)$. We denote by $\widetilde{\Gamma}_k^* = \widetilde{\Gamma}^*$ the set of all C^0 equivalence classes of fibers of elements of $\widetilde{\Gamma}_k$ and their suspensions. Then, we have

$$\widetilde{\Gamma}_k^* = \bigcup_{p-n=k} \Gamma_{n,p}^*,$$

where $\Gamma_{n,p}^*$ is the set of C^0 equivalence classes in $\widetilde{\Gamma}_k^*$ of fibers of maps between manifolds of dimensions n and p, and each $\Gamma_{n,p}^*$ is an ascending set. Furthermore, $\widetilde{\Gamma}_k^*$ is closed under suspension. Then, it is clear that the set of all $\widetilde{\Gamma}_k^*$-maps contain the original set $\widetilde{\Gamma}_k$ of maps. For a stable system of admissible equivalence relations $\mathcal{R}_k^{\widetilde{\Gamma}}$ among the elements of $\widetilde{\Gamma}_k^*$, we define the *universal complex of singular fibers for* $\widetilde{\Gamma}_k$ *with respect to* $\mathcal{R}_k^{\widetilde{\Gamma}}$ by

$$\mathcal{C}(\widetilde{\Gamma}_k, \mathcal{R}_k^{\widetilde{\Gamma}}) = \mathcal{C}(\widetilde{\Gamma}_k^*, \mathcal{R}_k^{\widetilde{\Gamma}}).$$

Furthermore, we denote the corresponding cohomology group of codimension κ by $H^\kappa(\widetilde{\Gamma}_k, \mathcal{R}_k^{\widetilde{\Gamma}})$. We call the set of $\widetilde{\Gamma}_k^*$-maps the *completion* of $\widetilde{\Gamma}_k$. When the set of $\widetilde{\Gamma}_k^*$-maps coincides with the original set $\widetilde{\Gamma}_k$, we say that the set $\widetilde{\Gamma}_k$ is *complete*.

Example 8.22. For example, the set $\mathcal{S}_{\rm pr}^0(n,p)$ is not complete, since there exist nonstable Thom maps whose fibers are all C^0 equivalent to a fiber of a C^0 stable map. On the other hand, $\mathcal{T}_{\rm pr}(n,p)$ is clearly complete.

In the following, if $\Gamma = \Gamma_{n,p} \subset \Gamma'_{n,p} = \Gamma' \subset \mathcal{T}_{\rm pr}(n,p)$ and ϱ^Γ is the restriction of an admissible equivalence relation $\varrho^{\Gamma'}$ for the fibers of elements of $\Gamma'_{n,p}$, we sometimes write $\mathcal{C}(\Gamma_{n,p}, \varrho^{\Gamma'})$ in place of $\mathcal{C}(\Gamma_{n,p}, \varrho^\Gamma)$ when there is no confusion. For the universal complexes for the fibers of codimension k maps, we sometimes use the same convention as well.

Example 8.23. Let $\mathcal{M}_{\rm pr}(n,p)$ be the set of all proper Morin maps in $\mathcal{T}_{\rm pr}(n,p)$ which satisfy the normal crossing condition as in [16, Chapter VI, §5], and set

$$\widetilde{\mathcal{M}}_{\rm pr}(k) = \bigcup_{p-n=k} \mathcal{M}_{\rm pr}(n,p).$$

(Here, a smooth map is called a *Morin map* if its singularities are all of Morin types [34].) Furthermore, we denote by $\mathcal{M}_{\rm pr}(n,p)^{\rm ori}$ the subset of $\mathcal{M}_{\rm pr}(n,p)$ consisting of those maps whose source manifolds are orientable, and we set

$$\widetilde{\mathcal{M}}_{\rm pr}(k)^{\rm ori} = \bigcup_{p-n=k} \mathcal{M}_{\rm pr}(n,p)^{\rm ori}.$$

(Note that the sets $\widetilde{\mathcal{M}}_{\rm pr}(k)$ and $\widetilde{\mathcal{M}}_{\rm pr}(k)^{\rm ori}$ are closed under suspension.) Then, by using Remark 8.20, we can show that

$$C^\kappa(\widetilde{\mathcal{M}}_{\mathrm{pr}}(-1)^{\mathrm{ori}}, \mathcal{R}^0_{-1}) = C^\kappa(\mathcal{M}_{\mathrm{pr}}(4,3)^{\mathrm{ori}}, \varrho^0_{4,3})$$

for all $\kappa \leq 3$, and hence

$$H^\kappa(\widetilde{\mathcal{M}}_{\mathrm{pr}}(-1)^{\mathrm{ori}}, \mathcal{R}^0_{-1}) = H^\kappa(\mathcal{M}_{\mathrm{pr}}(4,3)^{\mathrm{ori}}, \varrho^0_{4,3})$$

for $\kappa \leq 2$ (see also the paragraph just after Problem 8.17). Compare this with Problem 9.8.

9

Stable Maps of 4-Manifolds into 3-Manifolds

Now let us consider a more specific situation, i.e., the case of proper C^∞ stable maps of orientable 4-manifolds into 3-manifolds. Recall that a proper smooth map of a 4-manifold into a 3-manifold is C^∞ stable if and only if it is C^0 stable, as we have noted in Remark 3.2. In the following, we denote by $\mathcal{S}^0_{\mathrm{pr}}(n,p)^{\mathrm{ori}}$ the subset of $\mathcal{S}^0_{\mathrm{pr}}(n,p)$ consisting of the proper C^0 stable maps of *orientable* manifolds of dimension n into manifolds of dimension p.

The universal complex of singular fibers $\mathcal{C}(\mathcal{S}^0_{\mathrm{pr}}(4,3)^{\mathrm{ori}}, \varrho^0_{4,3})$ for proper C^0 stable maps of orientable 4-manifolds into 3-manifolds with respect to the C^0 (or C^∞) right-left equivalence $\varrho^0_{4,3}$ can be described as follows.

For a positive integer ℓ, let us denote by I^0_ℓ the equivalence class of the singular fiber which is the disjoint union of the corresponding singular fiber I^0 as in Fig. 3.4 and some fibers of the trivial circle bundle such that its total number of connected components is equal to ℓ. We define I^1_ℓ ($\ell \geq 1$), II^{00}_ℓ ($\ell \geq 2$), etc. similarly. Furthermore, let $\mathbf{0}_\ell$ ($\ell \geq 0$) denote the equivalence class of the regular fiber consisting of ℓ copies of a fiber of the trivial circle bundle.

Then, by the construction in Chap. 8, we obtain the complex of $\mathbf{Z}_2$-coefficients

$$\begin{aligned} 0 \longrightarrow C^0(\mathcal{S}^0_{\mathrm{pr}}(4,3)^{\mathrm{ori}}, \varrho^0_{4,3}) &\xrightarrow{\delta_0} C^1(\mathcal{S}^0_{\mathrm{pr}}(4,3)^{\mathrm{ori}}, \varrho^0_{4,3}) \\ &\xrightarrow{\delta_1} C^2(\mathcal{S}^0_{\mathrm{pr}}(4,3)^{\mathrm{ori}}, \varrho^0_{4,3}) \xrightarrow{\delta_2} C^3(\mathcal{S}^0_{\mathrm{pr}}(4,3)^{\mathrm{ori}}, \varrho^0_{4,3}) \longrightarrow 0, \end{aligned}$$

where $C^0(\mathcal{S}^0_{\mathrm{pr}}(4,3)^{\mathrm{ori}}, \varrho^0_{4,3})$ is generated by $\mathbf{0}_\ell$, and $C^i(\mathcal{S}^0_{\mathrm{pr}}(4,3)^{\mathrm{ori}}, \varrho^0_{4,3})$, $i = 1,2,3$, are generated by $\mathrm{I}^*_\ell, \mathrm{II}^*_\ell$ and III^*_ℓ, respectively, for various ℓ. Note that we have not specified any proper C^0 stable map $f : M \to N$ of an orientable 4-manifold into a 3-manifold. Hence, this complex can be regarded as a *universal complex of singular fibers for proper C^0 stable maps of orientable 4-manifolds into 3-manifolds*, in the sense that the corresponding complex for a specific C^0 stable map f is realized as a quotient complex of the universal complex (see Lemma 8.2).

This complex has the disadvantage that it has too many generators at each dimension and hence that it is a bit difficult to pursue a straightforward calculation of its cohomology groups. Thus, it seems reasonable to consider an equivalence relation weaker than the C^0 equivalence. For this, let us fix a positive integer m.

Definition 9.1. We say that two fibers of proper Thom maps between manifolds of dimensions $p+1$ and p, $p \geq 0$, are C^0 *equivalent modulo* m *circle components* if one of them is C^0 equivalent to the disjoint union of the other one and ℓm copies of a fiber of the trivial circle bundle for some nonnegative integer ℓ. We denote this equivalence relation by $\varrho^0_{p+1,p}(m)$. Given a subset $\Gamma_{p+1,p}$ of $\mathcal{T}_{\mathrm{pr}}(p+1,p)$, we shall use the same notation $\varrho^0_{p+1,p}(m)$ for the equivalence relation for $\Gamma^*_{p+1,p}$ induced by the above one, when there is no confusion (for the notation $\Gamma^*_{p+1,p}$, refer to Definition 8.21).

Lemma 9.2. *Let p be a nonnegative integer and m a positive integer. The C^0 equivalence modulo m circle components $\varrho^0_{p+1,p}(m)$ is an admissible equivalence relation for the fibers of elements of $\mathcal{T}_{\mathrm{pr}}(p+1,p)$ and hence for $\Gamma^*_{p+1,p}$.*

Proof. By definition, we see easily that condition (1) of Definition 7.2 is satisfied. Suppose that two fibers are C^0 equivalent modulo m circle components. Then by definition, the corresponding nearby fibers are all C^0 equivalent modulo m circle components. Hence condition (2) of Definition 7.2 is also satisfied. This completes the proof. □

Remark 9.3. Furthermore, we can also show that the system of admissible equivalence relations $\mathcal{R}^0_{-1}(m) = \{\varrho^0_{p+1,p}(m)\}_{p\geq 0}$ for the fibers of elements of $\widetilde{\mathcal{T}}_{\mathrm{pr}}(-1)$ is stable in the sense of Definition 8.5. Hence, for any subset $\widetilde{\Gamma}_{-1}$ of $\widetilde{\mathcal{T}}_{\mathrm{pr}}(-1)$ which is closed under suspension, the restriction of $\mathcal{R}^0_{-1}(m) = \{\varrho^0_{p+1,p}(m)\}_{p\geq 0}$ to $\widetilde{\Gamma}^*_{-1}$ is also stable.

By Lemma 9.2, for a nonnegative integer p and a positive integer m, we can define the universal complex of singular fibers

$$\mathcal{C}(\mathcal{T}_{\mathrm{pr}}(p+1,p), \varrho^0_{p+1,p}(m))$$

for proper Thom maps between manifolds of dimensions $p+1$ and p with respect to the C^0 equivalence modulo m circle components. More generally, for every subset $\Gamma_{p+1,p}$ of $\mathcal{T}_{\mathrm{pr}}(p+1,p)$, we can define the universal complex of singular fibers

$$\mathcal{C}(\Gamma_{p+1,p}, \varrho^0_{p+1,p}(m))$$

for $\Gamma_{p+1,p}$ with respect to the C^0 equivalence modulo m circle components (see Definition 8.21). We call the universal complexes thus obtained the *universal complexes of singular fibers modulo m circle components*.

The argument in Chap. 4 can be elaborated to prove the following results. Details are left to the reader.

Proposition 9.4. *The cohomology groups of the universal complex of singular fibers modulo two circle components*

$$\mathcal{C}(\mathcal{S}^0_{\mathrm{pr}}(4,3)^{\mathrm{ori}}, \varrho^0_{4,3}(2))$$

for proper C^0 *stable maps of orientable* 4*-manifolds into* 3*-manifolds are given as follows:*

$$H^0(\mathcal{S}^0_{\mathrm{pr}}(4,3)^{\mathrm{ori}}, \varrho^0_{4,3}(2)) \cong \mathbf{Z}_2 \ (\textit{generated by } [\mathbf{0}_{\mathrm{o}} + \mathbf{0}_{\mathrm{e}}]),$$
$$H^1(\mathcal{S}^0_{\mathrm{pr}}(4,3)^{\mathrm{ori}}, \varrho^0_{4,3}(2)) \cong \mathbf{Z}_2 \ (\textit{generated by } [\mathrm{I}^0_{\mathrm{o}} + \mathrm{I}^1_{\mathrm{e}}] = [\mathrm{I}^0_{\mathrm{e}} + \mathrm{I}^1_{\mathrm{o}}]),$$
$$H^2(\mathcal{S}^0_{\mathrm{pr}}(4,3)^{\mathrm{ori}}, \varrho^0_{4,3}(2)) = 0,$$

where $\mathfrak{F}_{\mathrm{o}}$ *(or* $\mathfrak{F}_{\mathrm{e}}$*) denotes the* C^0 *equivalence class modulo two circle components represented by* $\mathfrak{F}_\ell$ *with* ℓ *odd (resp. even), and* $[*]$ *denotes the cohomology class represented by the cocycle* $*$.

Remark 9.5. We can apply Proposition 7.4 as follows. The $\mathbf{Z}_2$-homology class (of closed support) in the target 3-manifold represented by a cycle corresponding to a coboundary of the universal complex of singular fibers (modulo m circle components) always vanishes. For $m = 2$, the coboundary groups are generated by the cochains listed in Table 9.1.

For $\kappa = 3$, we can easily read off the generators from Table 4.1 given in Chap. 4. Note that these lead to the congruences modulo two obtained in Proposition 4.1.

Compare this with [58, 12.5.4, 12.6.5, 13.4.1] and [38].

Remark 9.6. By using the classification theorems of singular fibers for proper C^0 stable maps in $\mathcal{S}^0_{\mathrm{pr}}(3,2)^{\mathrm{ori}}$ and in $\mathcal{S}^0_{\mathrm{pr}}(2,1)^{\mathrm{ori}}$ (see Remark 3.14 and Theorem 2.1), we see that the $\mathbf{Z}_2$-linear maps

$$s_\kappa : C^\kappa(\mathcal{S}^0_{\mathrm{pr}}(4,3)^{\mathrm{ori}}, \varrho^0_{4,3}(2)) \to C^\kappa(\mathcal{S}^0_{\mathrm{pr}}(3,2)^{\mathrm{ori}}, \varrho^0_{3,2}(2))$$

for $\kappa \leq 2$ and

$$s_\kappa : C^\kappa(\mathcal{S}^0_{\mathrm{pr}}(4,3)^{\mathrm{ori}}, \varrho^0_{4,3}(2)) \to C^\kappa(\mathcal{S}^0_{\mathrm{pr}}(2,1)^{\mathrm{ori}}, \varrho^0_{2,1}(2))$$

for $\kappa \leq 1$ induced by the suspension are in fact isomorphisms. Therefore, some of the above mentioned results are valid also for $\mathcal{C}(\mathcal{S}^0_{\mathrm{pr}}(3,2)^{\mathrm{ori}}, \varrho^0_{3,2}(2))$ and $\mathcal{C}(\mathcal{S}^0_{\mathrm{pr}}(2,1)^{\mathrm{ori}}, \varrho^0_{2,1}(2))$. For example, we have the following, where for an equivalence class of fibers, we use the same notation as the equivalence class of its suspension.

Table 9.1. Generators for the coboundary groups of $\mathcal{C}(\mathcal{S}^0_{\mathrm{pr}}(4,3)^{\mathrm{ori}}, \varrho^0_{4,3}(2))$

κ	generator(s)
1	$(\mathrm{I}^0_{\mathrm{e}} + \mathrm{I}^1_{\mathrm{o}}) - (\mathrm{I}^0_{\mathrm{o}} + \mathrm{I}^1_{\mathrm{e}})$
2	$\mathrm{II}^{01}_{\mathrm{o}} + \mathrm{II}^{01}_{\mathrm{e}} + \mathrm{II}^{a}_{\mathrm{e}}$, $\mathrm{II}^{01}_{\mathrm{o}} + \mathrm{II}^{01}_{\mathrm{e}} + \mathrm{II}^{a}_{\mathrm{o}}$

(1) For $f \in \mathcal{S}^0_{\mathrm{pr}}(3,2)^{\mathrm{ori}}$, we have

$$|\mathrm{II}^{01}_{\mathrm{o}}(f)|+|\mathrm{II}^{01}_{\mathrm{e}}(f)|+|\mathrm{II}^{a}_{\mathrm{e}}(f)| \equiv |\mathrm{II}^{01}_{\mathrm{o}}(f)|+|\mathrm{II}^{01}_{\mathrm{e}}(f)|+|\mathrm{II}^{a}_{\mathrm{o}}(f)| \equiv 0 \pmod 2.$$

(2) For $f \in \mathcal{S}^0_{\mathrm{pr}}(2,1)^{\mathrm{ori}}$, we have

$$|\mathrm{I}^0_{\mathrm{e}}(f)| + |\mathrm{I}^1_{\mathrm{o}}(f)| \equiv |\mathrm{I}^0_{\mathrm{o}}(f)| + |\mathrm{I}^1_{\mathrm{e}}(f)| \pmod 2.$$

We can also prove the following. For the notation, refer to Theorem 3.15 and Proposition 9.4.

Proposition 9.7. *The cohomology groups of the universal complex of singular fibers modulo two circle components*

$$\mathcal{C}(\mathcal{S}^0_{\mathrm{pr}}(3,2), \varrho^0_{3,2}(2))$$

for proper C^0 stable maps of (not necessarily orientable) 3-manifolds into surfaces are given as follows:

$$H^0(\mathcal{S}^0_{\mathrm{pr}}(3,2), \varrho^0_{3,2}(2)) \cong \mathbf{Z}_2 \ (\text{generated by } [\mathbf{0}_{\mathrm{o}} + \mathbf{0}_{\mathrm{e}}]),$$
$$H^1(\mathcal{S}^0_{\mathrm{pr}}(3,2), \varrho^0_{3,2}(2)) \cong \mathbf{Z}_2 \oplus \mathbf{Z}_2 \ (\text{generated by } [\widetilde{\mathrm{I}}^0_{\mathrm{o}} + \widetilde{\mathrm{I}}^1_{\mathrm{e}}] = [\widetilde{\mathrm{I}}^0_{\mathrm{e}} + \widetilde{\mathrm{I}}^1_{\mathrm{o}}] \text{ and } [\widetilde{\mathrm{I}}^2_{\mathrm{o}} + \widetilde{\mathrm{I}}^2_{\mathrm{e}}]).$$

The coboundary groups of the cochain complex $\mathcal{C}(\mathcal{S}^0_{\mathrm{pr}}(3,2), \varrho^0_{3,2}(2))$ are generated by the cochains listed in Table 9.2 (see also Remark 4.5).

By the same reason as in Remark 9.6, some of the above results hold also for $\mathcal{C}(\mathcal{S}^0_{\mathrm{pr}}(2,1), \varrho^0_{2,1}(2))$ as well.

Problem 9.8. Is the natural map

$$H^\kappa(\widetilde{\mathcal{S}}^0_{\mathrm{pr}}(-1), \mathcal{R}^0_{-1}) \to H^\kappa(\mathcal{S}^0_{\mathrm{pr}}(4,3), \varrho^0_{4,3})$$

an isomorphism for $\kappa \leq 2$? More generally, is the natural map

$$H^\kappa(\widetilde{\mathcal{S}}^0_{\mathrm{pr}}(-1), \mathcal{R}^0_{-1}) \to H^\kappa(\mathcal{S}^0_{\mathrm{pr}}(p+1,p), \varrho^0_{p+1,p})$$

an isomorphism for $\kappa \leq p-1$? Compare this with Remark 8.20 and Example 8.23.

Table 9.2. Generators for the coboundary groups of $\mathcal{C}(\mathcal{S}^0_{\mathrm{pr}}(3,2), \varrho^0_{3,2}(2))$

κ	generator(s)
1	$(\widetilde{\mathrm{I}}^0_{\mathrm{e}} + \widetilde{\mathrm{I}}^1_{\mathrm{o}}) - (\widetilde{\mathrm{I}}^0_{\mathrm{o}} + \widetilde{\mathrm{I}}^1_{\mathrm{e}})$
2	$\widetilde{\mathrm{II}}^{01}_{\mathrm{o}} + \widetilde{\mathrm{II}}^{01}_{\mathrm{e}} + \widetilde{\mathrm{II}}^{a}_{\mathrm{e}}$, $\widetilde{\mathrm{II}}^{01}_{\mathrm{o}} + \widetilde{\mathrm{II}}^{01}_{\mathrm{e}} + \widetilde{\mathrm{II}}^{a}_{\mathrm{o}}$, $\widetilde{\mathrm{II}}^{02}_{\mathrm{o}} + \widetilde{\mathrm{II}}^{02}_{\mathrm{e}} + \widetilde{\mathrm{II}}^{12}_{\mathrm{o}} + \widetilde{\mathrm{II}}^{12}_{\mathrm{e}} + \widetilde{\mathrm{II}}^{6}_{\mathrm{o}} + \widetilde{\mathrm{II}}^{6}_{\mathrm{e}}$

Now let us introduce the following equivalence relation among the fibers weaker than the C^0 equivalence.

Definition 9.9. Let us consider the class of maps f in $\mathcal{T}_{\mathrm{pr}}(n,p)$ such that the restriction to its singular set $S(f)$, $f|_{S(f)}$, is finite-to-one. We say that two fibers of such maps are C^0 *multi-singularity equivalent* (or *multi-singularity equivalent*) if the associated multi-germs at their singular points are C^0 right-left equivalent to each other. It is easy to show that this defines an admissible equivalence relation for the fibers of the above class of maps. Here, we adopt the convention that if the fibers contain no singular points, then they are always multi-singularity equivalent. We denote the multi-singularity equivalence relation by $\varrho^{\mathrm{ms}}_{n,p}$.

It is easy to see that if $n = p+1$, then

$$\varrho^{\mathrm{ms}}_{p+1,p} \leq \varrho^{0}_{p+1,p}(m) \leq \varrho^{0}_{p+1,p}$$

for every positive integer m.

Remark 9.10. Note that the universal complex of singular fibers with respect to the multi-singularity equivalence corresponds to Vassiliev's universal complex of multi-singularities [58] (see also [23, 38]).

By using a characterization of C^0 stable maps of orientable 5-dimensional manifolds into 4-dimensional manifolds as in Proposition 3.1, we can easily obtain the following. The details are left to the reader.

Proposition 9.11. *The cohomology groups of the universal complex of singular fibers for proper C^0 stable maps of orientable 5-dimensional manifolds into 4-dimensional manifolds with respect to the multi-singularity equivalence*

$$\mathcal{C}(\mathcal{S}^0_{\mathrm{pr}}(5,4)^{\mathrm{ori}}, \varrho^{\mathrm{ms}}_{5,4})$$

are given as follows:

$$H^0(\mathcal{S}^0_{\mathrm{pr}}(5,4)^{\mathrm{ori}}, \varrho^{\mathrm{ms}}_{5,4}) \cong \mathbf{Z}_2 \ (\textit{generated by } [\overline{\mathbf{0}}]),$$
$$H^1(\mathcal{S}^0_{\mathrm{pr}}(5,4)^{\mathrm{ori}}, \varrho^{\mathrm{ms}}_{5,4}) \cong \mathbf{Z}_2 \ (\textit{generated by } [\overline{\mathrm{I}}^0 + \overline{\mathrm{I}}^1]),$$
$$H^2(\mathcal{S}^0_{\mathrm{pr}}(5,4)^{\mathrm{ori}}, \varrho^{\mathrm{ms}}_{5,4}) = 0,$$
$$H^3(\mathcal{S}^0_{\mathrm{pr}}(5,4)^{\mathrm{ori}}, \varrho^{\mathrm{ms}}_{5,4}) = 0,$$

where $\overline{\mathbf{0}}$ denotes the multi-singularity equivalence class of regular fibers, $\overline{\mathrm{I}}^0$ the multi-singularity equivalence class of the definite fold mono-germ, $\overline{\mathrm{I}}^1$ the multi-singularity equivalence class of the indefinite fold mono-germ, and $[*]$ denotes the cohomology class represented by the cocycle $*$.

The above proposition shows that if we consider Vassiliev's universal complex of multi-singularities instead of our universal complex of singular fibers, then a result like Theorem 5.1 cannot be obtained. In fact, although we have not included the computation of $H^3(\mathcal{S}^0_{\mathrm{pr}}(5,4)^{\mathrm{ori}}, \varrho^0_{5,4}(2))$, we will see in Corollary 12.12 that it contains a nontrivial element which corresponds to the singular fiber of type III^8 as in Fig. 3.4 (see also Remark 10.12). We will also see that such a nontrivial element is closely related to the formula given in Theorem 5.1. This justifies our study of the universal complexes of singular fibers instead of multi-singularities.

10

Co-orientable Singular Fibers

Let us now proceed to the construction of another universal complex corresponding to co-orientable strata.

10.1 Complex with Respect to the C^0 Equivalence

Let us begin by the following definition.

Definition 10.1. Let $\mathfrak{F}$ be a C^0 equivalence class of fibers of proper Thom maps. Consider arbitrary homeomorphisms $\widetilde{\varphi}$ and φ which make the diagram

$$\begin{array}{ccc} (f^{-1}(U_0), f^{-1}(y)) & \xrightarrow{\widetilde{\varphi}} & (f^{-1}(U_1), f^{-1}(y)) \\ {\scriptstyle f}\downarrow & & \downarrow{\scriptstyle f} \\ (U_0, y) & \xrightarrow{\varphi} & (U_1, y) \end{array}$$

commutative, where f is a proper Thom map such that the fiber over y belongs to $\mathfrak{F}$, and U_i are open neighborhoods of y. Note that then we have $\varphi(\mathfrak{F}(f) \cap U_0) = \mathfrak{F}(f) \cap U_1$. We say that $\mathfrak{F}$ is *weakly co-orientable* if φ always preserves the local orientation of the normal bundle of $\mathfrak{F}(f)$ at y. We also call any fiber belonging to a weakly co-orientable C^0 equivalence class a *weakly co-orientable fiber*. In particular, if the codimension of $\mathfrak{F}$ coincides with the dimension of the target of f, then φ above should preserve the local orientation of the target at y.

Note that if $\mathfrak{F}$ is weakly co-orientable, then $\mathfrak{F}(f)$ has orientable normal bundle for every proper Thom map f. The author does not know whether the converse also holds or not.

Remark 10.2. Note that $\mathfrak{F}(f)$ is merely a C^0 submanifold of the target in general (see Lemma 7.1 and its proof) and we have to be careful when we talk about its normal bundle. However, as we have seen in the proof of Lemma 7.1, it is always locally flat and the orientability of its normal bundle is well-defined. For example, use the fact that $U_i \setminus (U_i \cap \mathfrak{F}(f))$ is homotopy equivalent to $S^{\kappa-1} \times (U_i \cap \mathfrak{F}(f))$ for appropriate U_i, where κ is the codimension of $\mathfrak{F}$.

Note that a weakly co-orientable C^0 equivalence class $\mathfrak{F}$ of fibers has exactly two co-orientations corresponding to the two orientations of the normal bundle of $\mathfrak{F}(f)$ at a point y in the target, where f is a Thom map such that the fiber over y belongs to $\mathfrak{F}$ and that the target itself is a small neighborhood of y. When one of the co-orientations is fixed, we call it a *co-oriented* C^0 equivalence class of fibers.

Using the co-orientations, we can construct the universal complex of weakly co-orientable singular fibers with coefficients in $\mathbf{Z}$ as follows. Let us first fix a dimension pair (n, p) with $p - n = k$. For $\kappa \in \mathbf{Z}$, let $CO^\kappa(\mathcal{T}_{\rm pr}(n, p), \varrho^0_{n,p})$ be the free $\mathbf{Z}$-module consisting of all formal linear combinations with integer coefficients, which may possibly contain infinitely many terms, of the C^0 equivalence classes $\mathfrak{F}$ of weakly co-orientable and co-oriented fibers of proper Thom maps between manifolds of dimensions n and p with $\kappa(\mathfrak{F}) = \kappa$, where $\varrho^0_{n,p}$ stands for the C^0 equivalence among the fibers of elements of $\mathcal{T}_{\rm pr}(n, p)$. In particular, $CO^\kappa(\mathcal{T}_{\rm pr}(n, p), \varrho^0_{n,p}) = 0$ for $\kappa > p$ and $\kappa < 0$. Here, we adopt the convention that -1 times a co-oriented C^0 equivalence class coincides with the C^0 equivalence class with reversed co-orientation. For two co-oriented C^0 equivalence classes of fibers $\mathfrak{F}$ and $\mathfrak{G}$ with $\kappa(\mathfrak{F}) = \kappa(\mathfrak{G}) - 1$, we define $[\mathfrak{F} : \mathfrak{G}] = n_{\mathfrak{F}}(\mathfrak{G}) \in \mathbf{Z}$, which is called the *incidence coefficient*, as in Chap. 7, where we take the co-orientations into account and the result is an integer.[1] Then we define the homomorphism

$$\delta_\kappa : CO^\kappa(\mathcal{T}_{\rm pr}(n, p), \varrho^0_{n,p}) \to CO^{\kappa+1}(\mathcal{T}_{\rm pr}(n, p), \varrho^0_{n,p})$$

by

$$\delta_\kappa(\mathfrak{F}) = \sum_{\kappa(\mathfrak{G})=\kappa+1} [\mathfrak{F} : \mathfrak{G}]\mathfrak{G}, \tag{10.1}$$

for $\mathfrak{F}$ with $\kappa(\mathfrak{F}) = \kappa$. Note that the homomorphism δ_κ is well-defined (for details, see the remarks just after (8.1) in Chap. 8).

Then, we can prove $\delta_{\kappa+1} \circ \delta_\kappa = 0$ as in Chap. 8 or in [58, §8]. Therefore,

$$\mathcal{CO}(\mathcal{T}_{\rm pr}(n, p), \varrho^0_{n,p}) = (CO^\kappa(\mathcal{T}_{\rm pr}(n, p), \varrho^0_{n,p}), \delta_\kappa)_\kappa$$

constitutes a complex and its cohomology groups

$$H^*(\mathcal{CO}(\mathcal{T}_{\rm pr}(n, p), \varrho^0_{n,p}))$$

are well-defined. We call the complex the *universal complex of weakly co-orientable singular fibers for proper Thom maps between manifolds of dimensions n and p with respect to the C^0 equivalence.*

Remark 10.3. Let $\mathfrak{F}$ be a weakly co-orientable and co-oriented C^0 equivalence class of fibers and $\mathfrak{G}$ a C^0 equivalence class of fibers such that $\kappa(\mathfrak{F}) = \kappa(\mathfrak{G}) - 1$.

[1] We can prove that the incidence coefficient in this sense is well-defined. Details are left to the reader.

Then, by using a "local co-orientation" for $\mathfrak{G}$, we can define the incidence coefficient $[\mathfrak{F} : \mathfrak{G}]$ as an integer. If this integer is non-zero, then we can show that $\mathfrak{G}$ is also weakly co-orientable. In other words, if $\mathfrak{G}$ is not weakly co-orientable, then the incidence coefficient $[\mathfrak{F} : \mathfrak{G}]$ necessarily vanishes.

By restricting the class of Thom maps that we consider, we can also obtain the universal complex of weakly co-orientable singular fibers for such a class of maps (for details, refer to §8.6). Such a complex is a quotient complex of the above constructed universal complex (see Lemma 8.19).

Example 10.4. For proper C^0 stable maps of orientable 4-manifolds into 3-manifolds, we see easily that $\mathbf{0}_\ell$, I^0_ℓ, I^1_ℓ, II^{01}_ℓ, II^a_ℓ, III^{0a}_ℓ, III^{1a}_ℓ, and III^b_ℓ are weakly co-orientable for every ℓ, and that the others are not weakly co-orientable. Using these weakly co-orientable fibers, we can construct the universal complex of weakly co-orientable singular fibers with coefficients in $\mathbf{Z}$ as follows:

$$0 \longrightarrow CO^0(\mathcal{S}^0_{\mathrm{pr}}(4,3)^{\mathrm{ori}}, \varrho^0_{4,3}) \xrightarrow{\delta_0} CO^1(\mathcal{S}^0_{\mathrm{pr}}(4,3)^{\mathrm{ori}}, \varrho^0_{4,3})$$
$$\xrightarrow{\delta_1} CO^2(\mathcal{S}^0_{\mathrm{pr}}(4,3)^{\mathrm{ori}}, \varrho^0_{4,3}) \xrightarrow{\delta_2} CO^3(\mathcal{S}^0_{\mathrm{pr}}(4,3)^{\mathrm{ori}}, \varrho^0_{4,3}) \longrightarrow 0.$$

We call this the *universal complex of weakly co-orientable singular fibers for proper* C^0 *stable maps of orientable* 4*-manifolds into* 3*-manifolds.*

By a method similar to that in §8.3, for an integer k, we can also define the universal complex $\mathcal{CO}(\widetilde{\mathcal{T}}_{\mathrm{pr}}(k), \mathcal{R}^0_k)$ of weakly co-orientable singular fibers for proper Thom maps of codimension k with respect to the stable system of C^0 equivalence relations as the projective limit of the complexes

$$\mathcal{CO}(\mathcal{T}_{\mathrm{pr}}(p-k,p), \varrho^0_{p-k,p}),$$

where $\mathcal{R}^0_k = \{\varrho^0_{p-k,p}\}_p$. We write the associated cohomology group of dimension κ by $H^\kappa(\mathcal{CO}(\widetilde{\mathcal{T}}_{\mathrm{pr}}(k), \mathcal{R}^0_k))$. As in §8.4, we can also give another description of this universal complex. Furthermore, we can also define a similar universal complex of weakly co-orientable singular fibers for a given class of Thom maps, and show that it is a quotient complex of the above constructed universal complex for Thom maps.

10.2 Complex with Respect to an Admissible Equivalence

Now let us fix an admissible equivalence relation $\varrho_{n,p}$ among the fibers of proper Thom maps between smooth manifolds of dimensions n and p. The following definition strongly depends on $\varrho_{n,p}$.

Definition 10.5. An equivalence class $\widetilde{\mathfrak{F}}$ of fibers of proper Thom maps with respect to $\varrho_{n,p}$ is *co-orientable* (or *strongly co-orientable*) if for any homeomorphism $\varphi : (U_0, y) \to (U_1, y)$ such that

$$\varphi(\widetilde{\mathfrak{G}}(f) \cap U_0) = \widetilde{\mathfrak{G}}(f) \cap U_1$$

for every equivalence class $\widetilde{\mathfrak{G}}$, φ preserves the local orientation of the normal bundle of $\widetilde{\mathfrak{F}}(f)$ at y, where f is a proper Thom map such that the fiber over y belongs to $\widetilde{\mathfrak{F}}$, and U_i are open neighborhoods of y. (Note that by Lemma 7.3, $\widetilde{\mathfrak{F}}(f)$ is a C^0 submanifold of the target.) In particular, if the codimension of $\widetilde{\mathfrak{F}}$ coincides with the dimension of the target, then φ should preserve the local orientation of the target at y. Note that if $\widetilde{\mathfrak{F}}$ is co-orientable, then $\widetilde{\mathfrak{F}}(f)$ has orientable normal bundle for every proper Thom map f, while the converse may not hold in general.

Remark 10.6. When the admissible equivalence relation $\varrho_{n,p}$ is given by the C^0 equivalence, i.e., when $\varrho_{n,p} = \varrho^0_{n,p}$, we can show that a C^0 equivalence class of fibers is weakly co-orientable if it is strongly co-orientable. The author does not know whether the converse also holds or not.

Using Definition 10.5, we can naturally define the *universal complex of co-orientable singular fibers for proper Thom maps between manifolds of dimensions n and p with respect to the admissible equivalence relation* $\varrho_{n,p}$, and we denote it by $\mathcal{CO}(\mathcal{T}_{\text{pr}}(n,p), \varrho_{n,p})$. Note that its cochain group at each dimension is a free $\mathbf{Z}$-module. We denote the corresponding cohomology group of dimension κ by $H^\kappa(\mathcal{CO}(\mathcal{T}_{\text{pr}}(n,p), \varrho_{n,p}))$.

If we are given a stable system of admissible equivalence relations

$$\mathcal{R}_k = \{\varrho_{p-k,p}\}_p$$

for the fibers of proper Thom maps of codimension k, then we can also define the corresponding universal complex $\mathcal{CO}(\widetilde{\mathcal{T}}_{\text{pr}}(k), \mathcal{R}_k)$ as the projective limit of the complexes $\mathcal{CO}(\mathcal{T}_{\text{pr}}(p-k,p), \varrho_{p-k,p})$ as in §8.3. We denote the corresponding cohomology group of dimension κ by $H^\kappa(\mathcal{CO}(\widetilde{\mathcal{T}}_{\text{pr}}(k), \mathcal{R}_k))$. Note that if the equivalence class of the suspension of a fiber whose equivalence class has codimension κ is co-orientable of codimension κ, then the original equivalence class is necessarily co-orientable, and hence the cochain map

$$\mathcal{CO}(\mathcal{T}_{\text{pr}}(n+\ell, p+\ell), \varrho_{n+\ell,p+\ell}) \to \mathcal{CO}(\mathcal{T}_{\text{pr}}(n,p), \varrho_{n,p})$$

is well-defined for every (n,p) with $p - n = k$ and $\ell > 0$. We can also give another description of the universal complex $\mathcal{CO}(\widetilde{\mathcal{T}}_{\text{pr}}(k), \mathcal{R}_k)$ as in §8.4.

Let us consider two admissible equivalence relations $\varrho_{n,p}$ and $\overline{\varrho}_{n,p}$ for the fibers of elements of $\mathcal{T}_{\text{pr}}(n,p)$. The following lemma is a direct consequence of Definition 10.5.

Lemma 10.7. *Suppose $\varrho_{n,p} \leq \overline{\varrho}_{n,p}$. Furthermore, suppose that $\widetilde{\mathfrak{F}}$ and $\overline{\mathfrak{F}}$ are equivalence classes with respect to $\varrho_{n,p}$ and $\overline{\varrho}_{n,p}$ respectively such that $\widetilde{\mathfrak{F}} \supset \overline{\mathfrak{F}}$ and that they have the same codimension. If $\widetilde{\mathfrak{F}}$ is co-orientable, then so is $\overline{\mathfrak{F}}$.*

By virtue of the above lemma, the homomorphism

$$\varepsilon_{\varrho_{n,p},\overline{\varrho}_{n,p}} : \mathcal{CO}(\mathcal{T}_{\mathrm{pr}}(n,p), \varrho_{n,p}) \to \mathcal{CO}(\mathcal{T}_{\mathrm{pr}}(n,p), \overline{\varrho}_{n,p})$$

as in (8.7) is a well-defined cochain map. We can also define a similar cochain map for two stable systems of admissible equivalence relations.

Furthermore, as in §8.6, by restricting the class of Thom maps that we consider, we can also obtain the universal complex of co-orientable singular fibers for such a class of maps. Such a complex is a quotient complex of one of the above constructed universal complexes of co-orientable singular fibers for proper Thom maps.

10.3 Stable Maps of 4-Manifolds into 3-Manifolds

Now let us consider proper C^0 stable maps of orientable 4-manifolds into 3-manifolds. By considering the C^0 equivalence modulo m circle components introduced in Definition 9.1, we get the corresponding universal complex of co-orientable singular fibers modulo m circle components. We denote the resulting complex by $\mathcal{CO}(\mathcal{S}^0_{\mathrm{pr}}(4,3)^{\mathrm{ori}}, \varrho^0_{4,3}(m))$. For $m = 2$, we see easily that $\mathbf{0}_\ell$, I^0_ℓ, I^1_ℓ, II^{01}_ℓ, II^a_ℓ, III^{0a}_ℓ, III^{1a}_ℓ, and III^b_ℓ are co-orientable for $\ell = \mathrm{o}, \mathrm{e}$, and that the others are not co-orientable, using the notation as in Proposition 9.4 (compare this with Example 10.4).

Then we easily get the following.[2]

Proposition 10.8. *The cohomology groups of the universal complex*

$$\mathcal{CO}(\mathcal{S}^0_{\mathrm{pr}}(4,3)^{\mathrm{ori}}, \varrho^0_{4,3}(2))$$

of co-orientable singular fibers modulo two circle components for proper C^0 stable maps of orientable 4-manifolds into 3-manifolds are given as follows:

$$H^0(\mathcal{CO}(\mathcal{S}^0_{\mathrm{pr}}(4,3)^{\mathrm{ori}}, \varrho^0_{4,3}(2))) \cong \mathbf{Z} \text{ (generated by } [\mathbf{0}_\mathrm{o} + \mathbf{0}_\mathrm{e}]),$$
$$H^1(\mathcal{CO}(\mathcal{S}^0_{\mathrm{pr}}(4,3)^{\mathrm{ori}}, \varrho^0_{4,3}(2))) \cong \mathbf{Z} \text{ (generated by } [\mathrm{I}^0_\mathrm{o} + \mathrm{I}^1_\mathrm{e}] = [\mathrm{I}^0_\mathrm{e} + \mathrm{I}^1_\mathrm{o}]),$$
$$H^2(\mathcal{CO}(\mathcal{S}^0_{\mathrm{pr}}(4,3)^{\mathrm{ori}}, \varrho^0_{4,3}(2))) = 0,$$

where $[*]$ denotes the cohomology class represented by the cocycle $*$.

---

[2] In what follows, for each co-orientable equivalence class, we fix its co-orientation once and for all, although we do not mention it explicitly.

Remark 10.9. As in Remark 9.5, if the target 3-manifold is orientable, then we can show that the integral homology class (of closed support) in the target 3-manifold represented by a cycle corresponding to a coboundary of the universal complex of co-orientable singular fibers (modulo m circle components) always vanishes. For $m = 2$, the coboundary groups are generated by the cochains listed in Table 10.1.

Thus, by the same reason as in Remark 9.6, we get the following proposition, where for an equivalence class of fibers, we use the same notation as the equivalence class of its suspension.

Proposition 10.10. (1) *Let $f : M \to N$ be a C^0 stable map of a closed orientable surface into a 1-dimensional manifold N. Then we have*

$$||\mathrm{I}^0_{\mathrm{e}}(f)|| + ||\mathrm{I}^1_{\mathrm{o}}(f)|| = ||\mathrm{I}^0_{\mathrm{o}}(f)|| + ||\mathrm{I}^1_{\mathrm{e}}(f)||,$$

where for a co-oriented equivalence class $\widetilde{\mathfrak{F}}$ of fibers, $||\widetilde{\mathfrak{F}}(f)||$ denotes the algebraic number of fibers of f of type $\widetilde{\mathfrak{F}}$.

(2) *Let $f : M \to N$ be a C^0 stable map of a closed orientable 3-manifold into an orientable surface N. Then we have*

$$||\mathrm{II}^{01}_{\mathrm{o}}(f)|| + ||\mathrm{II}^{01}_{\mathrm{e}}(f)|| + ||\mathrm{II}^{a}_{\mathrm{e}}(f)|| = ||\mathrm{II}^{01}_{\mathrm{o}}(f)|| + ||\mathrm{II}^{01}_{\mathrm{e}}(f)|| + ||\mathrm{II}^{a}_{\mathrm{o}}(f)|| = 0.$$

(3) *Let $f : M \to N$ be a C^0 stable map of a closed orientable 4-manifold into an orientable 3-manifold N. Then we have*

$$||\mathrm{III}^{0a}_{\mathrm{o}}(f)|| + ||\mathrm{III}^{1a}_{\mathrm{e}}(f)|| + ||\mathrm{III}^{b}_{\mathrm{e}}(f)|| = ||\mathrm{III}^{0a}_{\mathrm{e}}(f)|| + ||\mathrm{III}^{1a}_{\mathrm{o}}(f)|| + ||\mathrm{III}^{b}_{\mathrm{o}}(f)|| = 0.$$

Let us pose the following problem concerning Chaps. 8, 9 and 10.

Problem 10.11. Let us consider the homology class in the target represented by a cycle corresponding to a cocycle of the universal complex of (co-orientable) singular fibers representing a nontrivial cohomology class of the complex. Can it be written as a polynomial of some characteristic classes as in [58, 23, 38]? Can we find such a polynomial which is universal in a certain sense, like Thom polynomials for singularities? (For Thom polynomials, see [3, 19] for example.)

Table 10.1. Generators for the coboundary groups of $\mathcal{CO}(\mathcal{S}^0_{\mathrm{pr}}(4,3)^{\mathrm{ori}}, \varrho^0_{4,3}(2))$

κ	generator(s)
1	$(\mathrm{I}^0_{\mathrm{e}} + \mathrm{I}^1_{\mathrm{o}}) - (\mathrm{I}^0_{\mathrm{o}} + \mathrm{I}^1_{\mathrm{e}})$
2	$\mathrm{II}^{01}_{\mathrm{o}} + \mathrm{II}^{01}_{\mathrm{e}} + \mathrm{II}^{a}_{\mathrm{e}}$, $\mathrm{II}^{01}_{\mathrm{o}} + \mathrm{II}^{01}_{\mathrm{e}} + \mathrm{II}^{a}_{\mathrm{o}}$
3	$\mathrm{III}^{0a}_{\mathrm{o}} + \mathrm{III}^{1a}_{\mathrm{e}} + \mathrm{III}^{b}_{\mathrm{e}}$, $\mathrm{III}^{0a}_{\mathrm{e}} + \mathrm{III}^{1a}_{\mathrm{o}} + \mathrm{III}^{b}_{\mathrm{o}}$

Note that the above problem is closely related to the homomorphism which will be defined in Chap. 11.

Remark 10.12. We have not included the calculation of the third cohomology groups of the universal complexes of (co-orientable) singular fibers modulo two circle components for proper C^0 stable maps of orientable 4-manifolds into 3-manifolds, since the corresponding complex terminates essentially at dimension three. In order to calculate the third cohomology groups which make sense, we have to calculate the third cohomology group of the complex $\mathcal{C}(\mathcal{S}^0_{\mathrm{pr}}(5,4)^{\mathrm{ori}}, \varrho^0_{5,4}(2))$ (or $\mathcal{CO}(\mathcal{S}^0_{\mathrm{pr}}(5,4)^{\mathrm{ori}}, \varrho^0_{5,4}(2))$). In other words, we have to classify the singular fibers of proper C^0 stable maps of orientable 5-manifolds into 4-manifolds.

Nevertheless, Theorem 5.1 indicates that the singular fiber III^8 might represent a generator of the third cohomology group and that the corresponding homology class for proper C^0 stable maps of orientable 4-manifolds into 3-manifolds can be written in terms of a polynomial of Stiefel-Whitney classes. In fact, this will be shown to be correct in Chap. 12 by using Theorem 5.1 (see Corollary 12.12).

In the case where the dimensions are smaller by two, we can check that the above expectations are affirmative as follows. As we have seen in Proposition 9.7, the singular fiber $\widetilde{\mathrm{I}}^2$ (or more precisely, the cocycle $\widetilde{\mathrm{I}}^2_{\mathrm{o}} + \widetilde{\mathrm{I}}^2_{\mathrm{e}}$) represents a generator of the first cohomology group of the universal complex $\mathcal{C}(\mathcal{S}^0_{\mathrm{pr}}(3,2), \varrho^0_{3,2}(2))$ of singular fibers for proper C^0 stable maps of (not necessarily orientable) 3-manifolds into surfaces with respect to the C^0 equivalence modulo two circle components. Furthermore, by Corollary 2.4, the corresponding 0-dimensional homology class for Morse functions on surfaces is nothing but the Euler characteristic modulo two of the source surface, which coincides with its top Stiefel-Whitney class (for a more precise argument, see Chap. 14).

The following problem is closely related to Problem 4.6. See also Problem 11.14.

Problem 10.13. For each generator of the cohomology groups of the universal complex of (co-orientable) singular fibers for proper C^0 stable maps of orientable 4-manifolds into 3-manifolds, does there exist a C^0 stable map of a closed orientable 4-manifold into a 3-manifold whose corresponding cycle represents a nonzero element in the homology of the target?

Compare the above problem with [58, §17].

11

Homomorphism Induced by a Thom Map

In this chapter, we show that one can obtain a lot of information on the cohomology groups of the universal complexes of singular fibers by using concrete examples of Thom maps.

Let $\Gamma = \Gamma_{n,p}$ be a subset of $\mathcal{T}_{\mathrm{pr}}(n,p)$ and $\varrho^\Gamma = \varrho^\Gamma_{n,p}$ an admissible equivalence relation among the fibers of elements of $\Gamma_{n,p}$.

Definition 11.1. Let

$$c = \sum_{\kappa(\widetilde{\mathfrak{F}})=\kappa} n_{\widetilde{\mathfrak{F}}} \widetilde{\mathfrak{F}}$$

be a κ-dimensional cochain of the complex $\mathcal{C}(\Gamma_{n,p}, \varrho^\Gamma_{n,p})$, where $n_{\widetilde{\mathfrak{F}}} \in \mathbf{Z}_2$. For a Thom map $f : M \to N$ which is an element of $\Gamma^* = \Gamma^*_{n,p}$, we define $c(f)$ to be the closure of the set of points $y \in N$ such that the fiber over y belongs to some $\widetilde{\mathfrak{F}}$ with $n_{\widetilde{\mathfrak{F}}} \neq 0$. If c is a cocycle, then $c(f)$ is a $\mathbf{Z}_2$-cycle of closed support of codimension κ of the target manifold N. If in addition, M is closed and $\kappa > 0$, then $c(f)$ is a $\mathbf{Z}_2$-cycle in the usual sense.

When c is a cocycle of the complex $\mathcal{CO}(\Gamma_{n,p}, \varrho^\Gamma_{n,p})$, $c(f)$ is naturally a $\mathbf{Z}$-cycle, provided that the target manifold N is oriented.

Lemma 11.2. *Suppose that c and c' are κ-dimensional cocycles of the complex $\mathcal{C}(\Gamma_{n,p}, \varrho^\Gamma_{n,p})$ which are cohomologous. Then $c(f)$ and $c'(f)$ are homologous in N for every $f \in \Gamma^*_{n,p}$.*

Proof. There exists a $(\kappa - 1)$-dimensional cochain d of the complex such that $c - c' = \delta_{\kappa-1} d$. Then we see easily that $c(f) - c'(f) = \partial d(f)$, where $d(f)$ is defined similarly. Hence the result follows. □

Note that a similar result holds also for cocycles of the universal complex of co-orientable singular fibers.

Definition 11.3. Let α be a κ-dimensional cohomology class of the complex $\mathcal{C}(\Gamma_{n,p}, \varrho^\Gamma_{n,p})$. For a proper Thom map $f : M \to N$ which is an element of $\Gamma^*_{n,p}$, we define $\alpha(f) \in H^c_{p-\kappa}(N; \mathbf{Z}_2)$ to be the homology class represented by the cycle $c(f)$ of closed support, where c is a cocycle representing α and $p = \dim N$. By Lemma 11.2, this is well-defined. When M is closed and $\kappa > 0$, we can also regard $\alpha(f)$ as an element of $H_{p-\kappa}(N; \mathbf{Z}_2)$.

Then we can define the map

$$\varphi_f : H^\kappa(\Gamma_{n,p}, \varrho^\Gamma_{n,p}) \to H^\kappa(N; \mathbf{Z}_2)$$

by $\varphi_f(\alpha) = \alpha(f)^*$, where $\alpha(f)^* \in H^\kappa(N; \mathbf{Z}_2)$ is the Poincaré dual to $\alpha(f) \in H^c_{p-\kappa}(N; \mathbf{Z}_2)$. This is clearly a homomorphism and we call it the *homomorphism induced by the Thom map f*. When M is closed and $\kappa > 0$, we can also regard φ_f as a homomorphism into the cohomology group $H^\kappa_c(N; \mathbf{Z}_2)$ of compact support.

When the target manifold N is oriented, we can define

$$\varphi_f : H^\kappa(\mathcal{CO}(\Gamma_{n,p}, \varrho^\Gamma_{n,p})) \to H^\kappa(N; \mathbf{Z})$$

similarly.

Suppose that $\varrho^\Gamma = \varrho^{\Gamma_{n,p}}_{n,p}$ and $\overline{\varrho}^\Gamma = \overline{\varrho}^{\Gamma_{n,p}}_{n,p}$ are two admissible equivalence relations for the fibers of elements of $\Gamma = \Gamma_{n,p} \subset \mathcal{T}_{\mathrm{pr}}(n,p)$ such that $\varrho^{\Gamma_{n,p}}_{n,p} \leq \overline{\varrho}^{\Gamma_{n,p}}_{n,p}$. Then the following diagram is clearly commutative for every element $f : M \to N$ of Γ^*:

$$\begin{array}{ccc} H^\kappa(\Gamma, \varrho^\Gamma) & \xrightarrow{\varepsilon_{\varrho^\Gamma, \overline{\varrho}^\Gamma *}} & H^\kappa(\Gamma, \overline{\varrho}^\Gamma) \\ & {\scriptstyle \varphi_f}\searrow \quad \swarrow{\scriptstyle \varphi_f} & \\ & H^\kappa(N; \mathbf{Z}_2), & \end{array}$$

where $\varepsilon_{\varrho^\Gamma, \overline{\varrho}^\Gamma *} : H^\kappa(\Gamma, \varrho^\Gamma) \to H^\kappa(\Gamma, \overline{\varrho}^\Gamma)$ is the homomorphism induced by the natural cochain map $\varepsilon_{\varrho^\Gamma, \overline{\varrho}^\Gamma} : \mathcal{C}(\Gamma, \varrho^\Gamma) \to \mathcal{C}(\Gamma, \overline{\varrho}^\Gamma)$ defined in §§8.5 and 8.6.

Furthermore, if $\Gamma \subset \Gamma' \subset \mathcal{T}_{\mathrm{pr}}(n,p)$, then for every element $f : M \to N$ of Γ^*, we have the commutative diagram

$$\begin{array}{ccc} H^\kappa(\Gamma', \varrho^{\Gamma'}) & \xrightarrow{\pi_{\Gamma', \Gamma *}} & H^\kappa(\Gamma, \varrho^\Gamma) \\ & {\scriptstyle \varphi_f}\searrow \quad \swarrow{\scriptstyle \varphi_f} & \\ & H^\kappa(N; \mathbf{Z}_2), & \end{array}$$

where $\varrho^{\Gamma'}$ is an admissible equivalence relation among the fibers of elements of Γ', ϱ^Γ is its restriction to the fibers of elements of Γ, and $\pi_{\Gamma', \Gamma *}$ is the homomorphism induced by the natural cochain map $\pi_{\Gamma', \Gamma} : \mathcal{C}(\Gamma', \varrho^{\Gamma'}) \to \mathcal{C}(\Gamma, \varrho^\Gamma)$ defined in §8.6.

In particular, we have the commutative diagram

$$\begin{array}{ccccc} H^\kappa(\Gamma_{n,p}, \varrho^\Gamma_{n,p}) & & \to & & H^\kappa(\Gamma_{n,p}, \varrho^0_{n,p}) \\ & {\scriptstyle \varphi_f}\searrow & & \swarrow{\scriptstyle \varphi_f} & \\ \downarrow & & H^\kappa(N; \mathbf{Z}_2) & & \downarrow \\ & {\scriptstyle \varphi_f}\nearrow & & \nwarrow{\scriptstyle \varphi_f} & \\ H^\kappa(f, \varrho^\Gamma_{n,p}) & & \to & & H^\kappa(f, \varrho^0_{n,p}) \end{array}$$

for every element $f : M \to N$ of $\Gamma^*_{n,p}$, where $\Gamma_{n,p} \subset \mathcal{T}_{\mathrm{pr}}(n,p)$, $\varrho^{\Gamma}_{n,p}$ is an admissible equivalence relation for the fibers of elements of $\Gamma_{n,p}$, $\varrho^0_{n,p}$ denotes the C^0 equivalence, and the vertical and the horizontal homomorphisms are the natural ones defined as above (see also (8.8)).

Now as in §8.6, let

$$\widetilde{\Gamma} = \widetilde{\Gamma}_k = \bigcup_{p-n=k} \Gamma_{n,p}$$

be a set of C^0 equivalence classes of fibers of proper Thom maps of codimension k such that each $\Gamma_{n,p}$ is an ascending set of C^0 equivalence classes of fibers of elements of $\mathcal{T}_{\mathrm{pr}}(n,p)$, and that $\widetilde{\Gamma}$ is closed under suspension in the sense of Definition 8.4. Furthermore, let $\mathcal{R}^{\widetilde{\Gamma}}_k = \{\varrho^{\Gamma_{p-k,p}}_{p-k,p}\}_p$ be a system of equivalence relations, where each $\varrho^{\Gamma_{p-k,p}}_{p-k,p}$ is an admissible equivalence relation among the fibers of $\Gamma_{p-k,p}$-maps. We assume that the system $\mathcal{R}^{\widetilde{\Gamma}}_k$ of admissible equivalence relations is stable in the sense of Definition 8.5. Then, for every $f : M \to N$ in $\Gamma^*_{n,p}$ with $p-n=k$, we have the natural and well-defined homomorphism

$$\widetilde{\varphi}_f : H^\kappa(\widetilde{\Gamma}_k, \mathcal{R}^{\widetilde{\Gamma}}_k) \to H^\kappa(N; \mathbf{Z}_2)$$

which is defined as the composition of the homomorphism

$$\Phi^\kappa_{n,p*} : H^\kappa(\widetilde{\Gamma}_k, \mathcal{R}^{\widetilde{\Gamma}}_k) \to H^\kappa(\Gamma_{n,p}, \varrho^{\Gamma_{n,p}}_{n,p})$$

induced by the cochain map $\Phi^\kappa_{n,p} : C^\kappa(\widetilde{\Gamma}_k, \mathcal{R}^{\widetilde{\Gamma}}_k) \to C^\kappa(\Gamma_{n,p}, \varrho^{\Gamma_{n,p}}_{n,p})$ as in Remark 8.20 and the homomorphism

$$\varphi_f : H^\kappa(\Gamma_{n,p}, \varrho^{\Gamma_{n,p}}_{n,p}) \to H^\kappa(N; \mathbf{Z}_2)$$

defined above. We can also use the other description of the universal complex as in §8.4 and the definition as in Definition 11.3 in order to define $\widetilde{\varphi}_f$.

Remark 11.4. In the above situation, it is easy to verify that for every $f : M \to N$ in $\Gamma^*_{n,p}$ and a positive integer ℓ, the following diagram is commutative:

$$\begin{array}{ccc} H^\kappa(\Gamma_{n+\ell,p+\ell}, \varrho^{\Gamma_{n+\ell,p+\ell}}_{n+\ell,p+\ell}) & \xrightarrow{s_{\kappa*}} & H^\kappa(\Gamma_{n,p}, \varrho^{\Gamma_{n,p}}_{n,p}) \\ \downarrow{\scriptstyle \varphi_{\widetilde{f}}} & & \downarrow{\scriptstyle \varphi_f} \\ H^\kappa(N \times \mathbf{R}^\ell; \mathbf{Z}_2) & \cong & H^\kappa(N; \mathbf{Z}_2), \end{array}$$

where $s_{\kappa*}$ is the homomorphism induced by the suspension, $\widetilde{f} : M \times \mathbf{R}^\ell \to N \times \mathbf{R}^\ell$ is the ℓ-th suspension of f, and the last horizontal isomorphism is the natural one.

As a direct consequence of the above definitions, we have the following.

Proposition 11.5. *In the above situations, we have*

$$\mathrm{rank}_{\mathbf{Z}_2}\, \varphi_f \leq \dim_{\mathbf{Z}_2} H^\kappa(\Gamma_{n,p}, \varrho_{n,p}^{\Gamma_{n,p}})$$

and

$$\mathrm{rank}_{\mathbf{Z}_2}\, \widetilde{\varphi}_f \leq \dim_{\mathbf{Z}_2} H^\kappa(\widetilde{\Gamma}_k, \mathcal{R}_k^{\widetilde{\Gamma}}).$$

Remark 11.6. It is sometimes difficult to directly calculate the cohomology group $H^*(\Gamma_{n,p}, \varrho^0_{n,p})$ with respect to the C^0 equivalence. However, the above argument shows that if we have an element α of the cohomology group $H^*(\Gamma_{n,p}, \varrho_{n,p}^{\Gamma_{n,p}})$ with respect to an admissible equivalence relation $\varrho_{n,p}^{\Gamma_{n,p}}$ such that $\varphi_f(\alpha) \neq 0$ for some $f \in \Gamma^*_{n,p}$, then the image of α in $H^*(\Gamma_{n,p}, \varrho^0_{n,p})$ does not vanish. In other words, by calculating the cohomology group with respect to an admissible equivalence relation, which is often much easier than that with respect to the C^0 equivalence, and by constructing explicit examples, we can find nontrivial elements of the cohomology group with respect to the C^0 equivalence. This justifies our study developed in Chaps. 6 and 9.

Let us prepare some lemmas, which will be used later. For this, let us introduce the following definitions.

Definition 11.7. Let $f : M \to N$ be a proper Thom map and $g : V \to N$ a smooth map which is transverse to f and to all the strata of N. Put

$$\widetilde{V} = \{(x, y) \in M \times V \, : \, f(x) = g(y)\} \subset M \times V$$

and consider the following commutative diagram:

$$\begin{array}{ccc} \widetilde{V} & \xrightarrow{\ \widetilde{g}\ } & M \\ {\scriptstyle \widetilde{f}}\downarrow & & \downarrow{\scriptstyle f} \\ V & \xrightarrow{\ g\ } & N, \end{array}$$

where $\widetilde{g}$ and $\widetilde{f}$ are the restrictions of the projections to the first and the second factors respectively. Note that $\widetilde{V}$ is a smooth manifold of dimension $\dim V + \dim M - \dim N$ and that $\widetilde{f}$ is a proper Thom map. We call $\widetilde{f}$ the *pull-back of* f *by* g and say that $\widetilde{f}$ *is obtained by pulling back* f *by* g.

Definition 11.8. Suppose that

$$\Gamma_{n,p} \subset \mathcal{T}_{\mathrm{pr}}(n,p) \quad \text{and} \quad \Gamma_{n+\ell,p+\ell} \subset \mathcal{T}_{\mathrm{pr}}(n+\ell, p+\ell)$$

are given with $\ell > 0$ such that the ℓ-th suspension of an element of $\Gamma_{n,p}$ always belong to $\Gamma_{n+\ell,p+\ell}$. Let $f : M \to N$ be an arbitrary element of $\Gamma_{n+\ell,p+\ell}$ and $g : \mathrm{Int}\, D^p \to N$ an arbitrary smooth map which is transverse to f and to all the strata of N. Note that the pull-back $\widetilde{f}$ of f by g is then an element of $\mathcal{T}_{\mathrm{pr}}(n,p)$. If the fibers of $\widetilde{f}$ always belong to $\Gamma^*_{n,p}$, then we say that $\Gamma_{n,p}$ is *transversely complete with respect to* $\Gamma_{n+\ell,p+\ell}$.

Furthermore, we say that

$$\widetilde{\Gamma}_k = \bigcup_{p-n=k} \Gamma_{n,p} \subset \widetilde{\mathcal{T}}_{\rm pr}(k)$$

is *transversely complete* if it is closed under suspension and if $\Gamma_{n,p}$ is transversely complete with respect to $\Gamma_{n+\ell,p+\ell}$ for all n, p and ℓ.

Note that the set $\widetilde{\mathcal{T}}_{\rm pr}(k)$ is clearly transversely complete.

The following lemma can be proved by the same argument as in the proof of Lemma 8.6. Details are left to the reader.

Lemma 11.9. *If $\Gamma_{n,p}$ is transversely complete with respect to $\Gamma_{n+\ell,p+\ell}$, then the natural $\mathbf{Z}_2$-linear map*

$$s_\kappa : C^\kappa(\Gamma_{n+\ell,p+\ell}, \varrho_{n+\ell,p+\ell}^{\Gamma_{n+\ell,p+\ell}}) \to C^\kappa(\Gamma_{n,p}, \varrho_{n,p}^{\Gamma_{n,p}})$$

induced by the suspension is a monomorphism for every $\kappa \leq p$, where $\varrho_{n+\ell,p+\ell}^{\Gamma_{n+\ell,p+\ell}}$ and $\varrho_{n,p}^{\Gamma_{n,p}}$ are admissible equivalence relations for the fibers of elements of $\Gamma_{n+\ell,p+\ell}$ and $\Gamma_{n,p}$, respectively, which are stable in a sense similar to Definition 8.5.

In the following lemma, we assume that each $\Gamma_{n,p}$, $p-n=k$, is a subset of $\mathcal{T}_{\rm pr}(n,p)$ and that $\widetilde{\Gamma}_k = \cup_{p-n=k}\Gamma_{n,p}$ is closed under suspension. Furthermore, $\{\varrho_{n,p}^{\Gamma_{n,p}}\}_{p-n=k}$ is a stable system of admissible equivalence relations for the fibers of elements of $\widetilde{\Gamma}_k$, where each $\varrho_{n,p}^{\Gamma_{n,p}}$ is an admissible equivalence relation for $\Gamma^*_{n,p}$. Recall that $\Gamma^*_{n,p}$ denotes the set of C^0 equivalence classes of fibers of elements of $\Gamma_{n,p}$ and, when there is no confusion, it also denotes the set of all $\Gamma^*_{n,p}$-maps (see §8.6).

Lemma 11.10. *Let $\alpha \in H^p(\Gamma_{n+\ell,p+\ell}, \varrho_{n+\ell,p+\ell}^{\Gamma_{n+\ell,p+\ell}})$ be a cohomology class such that $\varphi_f(s_{p*}\alpha) = 0$ in $H^p(N;\mathbf{Z}_2)$ for every Thom map $f : M \to N$ in $\Gamma^*_{n,p}$ with both M and N being closed, where*

$$s_{p*} : H^p(\Gamma_{n+\ell,p+\ell}, \varrho_{n+\ell,p+\ell}^{\Gamma_{n+\ell,p+\ell}}) \to H^p(\Gamma_{n,p}, \varrho_{n,p}^{\Gamma_{n,p}})$$

is the homomorphism induced by the suspension, and

$$\varphi_f : H^p(\Gamma_{n,p}, \varrho_{n,p}^{\Gamma_{n,p}}) \to H^p(N;\mathbf{Z}_2)$$

is the homomorphism induced by f. Furthermore, suppose that $\Gamma_{n,p}$ is transversely complete with respect to $\Gamma_{n+\ell,p+\ell}$. Then, for every $g : M' \to N'$ in $\Gamma^*_{n+\ell,p+\ell}$, we have $\varphi_g(\alpha) = 0$ in $H^p(N';\mathbf{Z}_2)$.

Proof. Let c be a cocycle of $C^p(\Gamma_{n+\ell,p+\ell}, \varrho_{n+\ell,p+\ell}^{\Gamma_{n+\ell,p+\ell}})$ which represents α. We have only to show that the homology class $[c(g)] \in H^c_\ell(N';\mathbf{Z}_2)$ represented by $c(g)$ vanishes. For this, it suffices to prove that the intersection number $[c(g)] \cdot \xi$ vanishes for all $\xi \in H_p(N'; \mathbf{Z}_2)$ by Poincaré duality.

By [56], there exists a closed p-dimensional manifold V and a smooth map $h : V \to N'$ such that $h_*[V]_2 = \xi$, where $[V]_2 \in H_p(V; \mathbf{Z}_2)$ is the fundamental class of V. We may assume that h is transverse to g and to all the strata of N'. Let us consider the pull-back $\widetilde{g} : \widetilde{V} \to V$ of g by h (see Definition 11.7). Since $\Gamma_{n,p}$ is transversely complete with respect to $\Gamma_{n+\ell,p+\ell}$ by our assumption, we see that $\widetilde{g}$ is an element of $\Gamma^*_{n,p}$. Furthermore, both the source and the target manifolds of $\widetilde{g}$ are closed. Therefore, by our assumption, $\varphi_{\widetilde{g}}(s_{p*}\alpha) = 0$; in other words, $(s_p c)(\widetilde{g})$ consists of an even number of points in V. Hence, the intersection number of $[c(g)]$ and ξ vanishes. This completes the proof. □

In fact, we have the following.

Lemma 11.11. *Suppose that $\Gamma_{n,p}$ is transversely complete with respect to $\Gamma_{n+\ell,p+\ell}$. Then, for a cohomology class $\alpha \in H^p(\Gamma_{n+\ell,p+\ell}, \varrho_{n+\ell,p+\ell}^{\Gamma_{n+\ell,p+\ell}})$, the following two are equivalent to each other.*

(1) *For every proper Thom map $f : M \to N$ in $\Gamma^*_{n,p}$, we have $\varphi_f(s_{p*}\alpha) = 0$ in $H^p(N; \mathbf{Z}_2)$.*
(2) *For every proper Thom map $g : M' \to N'$ in $\Gamma^*_{n+\ell,p+\ell}$, we have $\varphi_g(\alpha) = 0$ in $H^p(N'; \mathbf{Z}_2)$.*

Proof. We have already proved that (1) implies (2). Suppose that (2) holds. For a given proper Thom map $f : M \to N$ in $\Gamma^*_{n,p}$, consider the commutative diagram given in Remark 11.4 with $\kappa = p$. Since the ℓ-th suspension $\widetilde{f} : M \times \mathbf{R}^\ell \to N \times \mathbf{R}^\ell$ is a $\Gamma^*_{n+\ell,p+\ell}$-map, we have $\varphi_{\widetilde{f}}(\alpha) = 0$ by our assumption. Hence (1) follows. □

Remark 11.12. In fact, we can prove the following, without assuming that $\Gamma_{n,p}$ is transversely complete with respect to $\Gamma_{n+\ell,p+\ell}$. For a cohomology class $\beta \in \operatorname{Im} s_{p*}$, the following two are equivalent to each other, where

$$s_{p*} : H^p(\Gamma_{n+\ell,p+\ell}, \varrho_{n+\ell,p+\ell}^{\Gamma_{n+\ell,p+\ell}}) \to H^p(\Gamma_{n,p}, \varrho_{n,p}^{\Gamma_{n,p}})$$

is the homomorphism induced by the suspension.

(1) For every proper Thom map $f : M \to N$ in $\Gamma^*_{n,p}$, we have $\varphi_f(\beta) = 0$ in $H^p(N; \mathbf{Z}_2)$.
(2) For every Thom map $f : M \to N$ in $\Gamma^*_{n,p}$ with both M and N being closed, we have $\varphi_f(\beta) = 0$ in $H^p(N; \mathbf{Z}_2)$.

The proof goes as follows. Suppose (2) holds and take f as in (1). Consider the commutative diagram given in Remark 11.4 with $\kappa = p$. Then apply an argument similar to that in the proof of Lemma 11.10, using a smooth map h of a closed p-dimensional manifold into $N \times \mathbf{R}^\ell$. Since $\widetilde{f}$ is the ℓ-th suspension of f, the pull-back of $\widetilde{f}$ by h is a $\Gamma^*_{n,p}$-map. Hence, from (2), (1) follows.

Note that all the results in this chapter hold also for the universal complexes of co-orientable singular fibers and the cohomology groups with **Z**-coefficients, provided that the target manifolds are oriented.

Problem 11.13. Are the cohomology groups

$$H^\kappa(\mathcal{T}_{\rm pr}(n,p), \varrho^0_{n,p}), \tag{11.1}$$
$$H^\kappa(\mathcal{S}^0_{\rm pr}(n,p), \varrho^0_{n,p}), \tag{11.2}$$
$$H^\kappa(\mathcal{CO}(\mathcal{T}_{\rm pr}(n,p), \varrho^0_{n,p})), \tag{11.3}$$
$$H^\kappa(\mathcal{CO}(\mathcal{S}^0_{\rm pr}(n,p), \varrho^0_{n,p})), \tag{11.4}$$
$$H^\kappa(\widetilde{\mathcal{T}}_{\rm pr}(k), \mathcal{R}^0_k), \tag{11.5}$$
$$H^\kappa(\widetilde{\mathcal{S}}^0_{\rm pr}(k), \mathcal{R}^0_k), \tag{11.6}$$
$$H^\kappa(\mathcal{CO}(\widetilde{\mathcal{T}}_{\rm pr}(k), \mathcal{R}^0_k)), \tag{11.7}$$
$$H^\kappa(\mathcal{CO}(\widetilde{\mathcal{S}}^0_{\rm pr}(k), \mathcal{R}^0_k)) \tag{11.8}$$

finitely generated for all κ?

The following is a generalization of Problem 10.13.

Problem 11.14. Let α be an element of one of the cohomology groups (11.1)–(11.8). If $\alpha \neq 0$, then does there exist a smooth map f (in the relevant class) such that $\varphi_f(\alpha)$ does not vanish? In other words, if $\varphi_f(\alpha) = 0$ for all f, then does α vanish?

As to an interpretation of the above problem, see Remark 12.14.

12

Cobordism Invariance

In this chapter, we define cobordisms of singular maps with a given set of singular fibers and show that the homomorphism φ_f induced by a Thom map f defined in Chap. 11 is a cobordism invariant of f when restricted to a certain subgroup. We also apply this notion of cobordisms to give a necessary and sufficient condition for a certain cochain of the universal complex to be a cocycle.

12.1 Cobordism of Singular Maps

As in §8.6, let

$$\widetilde{\Gamma} = \widetilde{\Gamma}_k = \bigcup_{p-n=k} \Gamma_{n,p}$$

be a set of C^0 equivalence classes of fibers of proper Thom maps of codimension k such that each $\Gamma_{n,p}$ is an ascending set of C^0 equivalence classes of fibers of elements of $\mathcal{T}_{\mathrm{pr}}(n,p)$, and that $\widetilde{\Gamma}$ is closed under suspension in the sense of Definition 8.4. Recall that a proper Thom map $f : M \to N$ of codimension k is a $\widetilde{\Gamma}_k$-map if its fibers all lie in $\widetilde{\Gamma}_k$. If M is a manifold with boundary, then we also suppose that $f(\partial M) \subset \partial N$ and for collar neighborhoods $C = \partial M \times [0,1)$ and $C' = \partial N \times [0,1)$ of ∂M and ∂N respectively, we have $f|_C = (f|_{\partial M}) \times \mathrm{id}_{[0,1)}$.

Definition 12.1. For a smooth manifold N, two $\widetilde{\Gamma}_k$-maps $f_0 : M_0 \to N$ and $f_1 : M_1 \to N$ of closed manifolds M_0 and M_1 are said to be *$\widetilde{\Gamma}_k$-cobordant* if there exist a compact manifold W with boundary the disjoint union of M_0 and M_1, and a $\widetilde{\Gamma}_k$-map $F : W \to N \times [0,1]$ such that $f_i = F|_{M_i} : M_i \to N \times \{i\}$, $i = 0, 1$. We call F a *$\widetilde{\Gamma}_k$-cobordism* between f_0 and f_1.

When M_i are oriented and W can be taken to be oriented so that $\partial W = (-M_0) \amalg M_1$, then we say that f_0 and f_1 are *oriented $\widetilde{\Gamma}_k$-cobordant.*

Remark 12.2. The notion of $\widetilde{\Gamma}_k$-maps and that of $\widetilde{\Gamma}_k$-cobordisms were essentially introduced by Rimányi and Szűcs [40], although they considered only the nonnegative codimension case and they called them τ-maps and τ-cobordisms respectively. Note that if the codimension is nonnegative, then a fiber of a proper generic map is always a finite set of points and that map-germs along the fibers are nothing but multi-germs. In the nonnegative codimension case, Rimányi and Szűcs constructed a universal $\widetilde{\Gamma}_k$-map and this gives rise to a lot of $\widetilde{\Gamma}_k$-cobordism invariants. Our aim in this chapter is to construct invariants of $\widetilde{\Gamma}_k$-cobordisms even in the negative codimension case.

Remark 12.3. In Definition 12.1, when the dimensions of the source manifolds M_0 and M_1 are equal to n, we have only to give $\Gamma_{n,p}$ and $\Gamma_{n+1,p+1}$ instead of the whole $\widetilde{\Gamma}_k$ in order to define the notion of $\widetilde{\Gamma}_k$-cobordisms. For this reason, we will sometimes talk about $\widetilde{\Gamma}_k$-cobordisms even when only $\Gamma_{n,p}$ and $\Gamma_{n+1,p+1}$ are given.

Let

$$s_{\kappa *} : H^{\kappa}(\Gamma_{n+1,p+1}, \varrho_{n+1,p+1}^{\Gamma_{n+1,p+1}}) \to H^{\kappa}(\Gamma_{n,p}, \varrho_{n,p}^{\Gamma_{n,p}})$$

be the homomorphism induced by the suspension, where $\mathcal{R}_k^{\widetilde{\Gamma}} = \{\varrho_{p-k,p}^{\Gamma_{p-k,p}}\}_p$ is a stable system of admissible equivalence relations for $\widetilde{\Gamma}$.

Lemma 12.4. *Let $f_i : M_i \to N$, $i = 0, 1$, be Thom maps which are elements of $\Gamma_{n,p}$ and are $\widetilde{\Gamma}_k$-maps, where we assume that M_i are closed. If they are $\widetilde{\Gamma}_k$-cobordant, then for every κ we have*

$$\varphi_{f_0}|_{\mathrm{Im}\, s_{\kappa *}} = \varphi_{f_1}|_{\mathrm{Im}\, s_{\kappa *}} : \mathrm{Im}\, s_{\kappa *} \to H^{\kappa}(N; \mathbf{Z}_2).$$

Proof. Let $F : W \to N \times [0,1]$ be a $\widetilde{\Gamma}_k$-cobordism between f_0 and f_1. Let c be an arbitrary κ-dimensional cocycle of the complex $\mathcal{C}(\Gamma_{n+1,p+1}, \varrho_{n+1,p+1}^{\Gamma_{n+1,p+1}})$ and set $\bar{c} = s_{\kappa}(c) \in C^{\kappa}(\Gamma_{n,p}, \varrho_{n,p}^{\Gamma_{n,p}})$. Then we see easily that $\partial c(F) = \bar{c}(f_1) \times \{1\} - \bar{c}(f_0) \times \{0\}$, since c is a cocycle (for the notation, refer to Definition 11.1). Then the result follows immediately. □

Remark 12.5. In Lemma 12.4, if $\kappa \geq 1$, then φ_{f_i} can be regarded as homomorphisms into $H_c^{\kappa}(N; \mathbf{Z}_2)$, since M_i are closed. In this case, we can prove that

$$\varphi_{f_0}|_{\mathrm{Im}\, s_{\kappa *}} = \varphi_{f_1}|_{\mathrm{Im}\, s_{\kappa *}} : \mathrm{Im}\, s_{\kappa *} \to H_c^{\kappa}(N; \mathbf{Z}_2).$$

Definition 12.6. The pairs $\{\Gamma_{n+1,p+1}, \varrho_{n+1,p+1}^{\Gamma_{n+1,p+1}}\}$ and $\{\Gamma_{n,p}, \varrho_{n,p}^{\Gamma_{n,p}}\}$ are said to be *compatible at dimension* κ if the homomorphism

$$s_{\kappa *} : H^{\kappa}(\Gamma_{n+1,p+1}, \varrho_{n+1,p+1}^{\Gamma_{n+1,p+1}}) \to H^{\kappa}(\Gamma_{n,p}, \varrho_{n,p}^{\Gamma_{n,p}})$$

is surjective.

Lemma 12.7. *The pairs* $\{\Gamma_{n+1,p+1}, \varrho_{n+1,p+1}^{\Gamma_{n+1,p+1}}\}$ *and* $\{\Gamma_{n,p}, \varrho_{n,p}^{\Gamma_{n,p}}\}$ *are compatible at dimension* κ *if the following conditions hold.*

(1) *Every fiber in* $\Gamma_{n+1,p+1}$ *of codimension* $\kappa+1$ *with respect to* $\varrho_{n+1,p+1}^{\Gamma_{n+1,p+1}}$ *is a suspension of a fiber in* $\Gamma_{n,p}$ *of the same codimension with respect to* $\varrho_{n,p}^{\Gamma_{n,p}}$.
(2) *If an equivalence class of fibers in* $\Gamma_{n,p}$ *with respect to* $\varrho_{n,p}^{\Gamma_{n,p}}$ *has codimension* κ*, then the equivalence class of their suspensions with respect to* $\varrho_{n+1,p+1}^{\Gamma_{n+1,p+1}}$ *has also codimension* κ.
(3) *Two fibers in* $\Gamma_{n,p}$ *whose equivalence classes with respect to* $\varrho_{n,p}^{\Gamma_{n,p}}$ *have codimension* κ *are equivalent with respect to* $\varrho_{n,p}^{\Gamma_{n,p}}$ *if and only if their suspensions are equivalent with respect to* $\varrho_{n+1,p+1}^{\Gamma_{n+1,p+1}}$.

Proof. If $\kappa < 0$ or $\kappa > p$, then the result is trivial. When $0 \leq \kappa \leq p$, by an argument similar to that in Remark 8.20, we see that if conditions (2) and (3) are satisfied, then

$$s_\kappa : C^\kappa(\Gamma_{n+1,p+1}, \varrho_{n+1,p+1}^{\Gamma_{n+1,p+1}}) \to C^\kappa(\Gamma_{n,p}, \varrho_{n,p}^{\Gamma_{n,p}})$$

is an epimorphism. Furthermore, condition (1) implies that $s_{\kappa+1}$ is a monomorphism (see also Remark 8.20). Thus the homomorphism $s_{\kappa *}$ induced on the κ-dimensional cohomology is an epimorphism, and hence the compatibility follows. □

Corollary 12.8. *Consider the case where* $\Gamma_{n,p} = \mathcal{T}_{\mathrm{pr}}(n,p)$ *for all* (n,p) *with* $p-n=k$*, and put* $\varrho_{n,p} = \varrho_{n,p}^{\Gamma_{n,p}}$*. We suppose that* $\kappa+1 \leq p$*. Then the pairs* $\{\mathcal{T}_{\mathrm{pr}}(n+1,p+1), \varrho_{n+1,p+1}\}$ *and* $\{\mathcal{T}_{\mathrm{pr}}(n,p), \varrho_{n,p}\}$ *are compatible at dimension* κ *if the following two conditions hold.*

(1) *If an equivalence class of fibers of elements of* $\mathcal{T}_{\mathrm{pr}}(n,p)$ *with respect to* $\varrho_{n,p}$ *has codimension* κ*, then the equivalence class of their suspensions with respect to* $\varrho_{n+1,p+1}$ *has also codimension* κ.
(2) *Two fibers of elements of* $\mathcal{T}_{\mathrm{pr}}(n,p)$ *whose equivalence classes with respect to* $\varrho_{n,p}$ *have codimension* κ *are equivalent with respect to* $\varrho_{n,p}$ *if and only if their suspensions are equivalent with respect to* $\varrho_{n+1,p+1}$.

Proof. Recall that if $\kappa+1 \leq p$, then by Lemma 8.6,

$$s_{\kappa+1} : C^{\kappa+1}(\mathcal{T}_{\mathrm{pr}}(n+1,p+1), \varrho_{n+1,p+1}) \to C^{\kappa+1}(\mathcal{T}_{\mathrm{pr}}(n,p), \varrho_{n,p})$$

is always a monomorphism. Then the result follows from Lemma 12.7. □

Corollary 12.9. *We suppose that the pairs* $\{\Gamma_{n+1,p+1}, \varrho_{n+1,p+1}^{\Gamma_{n+1,p+1}}\}$ *and* $\{\Gamma_{n,p}, \varrho_{n,p}^{\Gamma_{n,p}}\}$ *are compatible at dimension* κ*. Let* $f_i : M_i \to N$*,* $i = 0, 1$*, be Thom maps which are elements of* $\Gamma_{n,p}$ *and are* $\widetilde{\Gamma}_k$*-maps, where we assume that* M_i *are closed. If they are* $\widetilde{\Gamma}_k$*-cobordant, then we have*

$$\varphi_{f_0} = \varphi_{f_1} : H^\kappa(\Gamma_{n,p}, \varrho_{n,p}^{\Gamma_{n,p}}) \to H^\kappa(N; \mathbf{Z}_2).$$

If $\kappa \geq 1$, then we also have

$$\varphi_{f_0} = \varphi_{f_1} : H^\kappa(\Gamma_{n,p}, \varrho_{n,p}^{\Gamma_{n,p}}) \to H_c^\kappa(N; \mathbf{Z}_2).$$

By using a natural generalization of Proposition 8.15 to certain subsets of $\widetilde{\mathcal{T}}_{\mathrm{pr}}(k)$ together with an argument similar to that in the proof of Lemma 12.4, we get the following as well.

Corollary 12.10. *Let $\mathcal{R}_k^{\widetilde{\Gamma}} = \{\varrho_{p-k,p}^{\Gamma_{p-k,p}}\}_p$ be a stable system of admissible equivalence relations for $\widetilde{\Gamma}$. Let $f_i : M_i \to N$, $i = 0, 1$, be $\widetilde{\Gamma}$-maps with* $\dim M_i = n$ *and* $\dim N = p$, *where we assume that M_i are closed. If they are $\widetilde{\Gamma}$-cobordant, then for every κ we have*

$$\widetilde{\varphi}_{f_0} = \widetilde{\varphi}_{f_1} : H^\kappa(\widetilde{\Gamma}, \mathcal{R}_k^{\widetilde{\Gamma}}) \to H^\kappa(N; \mathbf{Z}_2).$$

If $\kappa \geq 1$, then we also have

$$\widetilde{\varphi}_{f_0} = \widetilde{\varphi}_{f_1} : H^\kappa(\widetilde{\Gamma}, \mathcal{R}_k^{\widetilde{\Gamma}}) \to H_c^\kappa(N; \mathbf{Z}_2).$$

When the manifold N is oriented, we get similar results in coefficients in $\mathbf{Z}$ by using the universal complex of co-orientable singular fibers. Details are left to the reader.

12.2 A Characterization of Cocycles

In this section, we shall give a necessary and sufficient condition for a certain cochain of the universal complex to be a cocycle in terms of the homomorphism induced by Thom maps.

Let $\widetilde{\Gamma} = \widetilde{\Gamma}_k$ be as in the previous section, and let $\mathcal{R}_k^{\widetilde{\Gamma}} = \{\varrho_{p-k,p}^{\Gamma_{p-k,p}}\}_p$ be a stable system of admissible equivalence relations for $\widetilde{\Gamma}$.

Let c be an arbitrary cochain in $C^\kappa(\Gamma_{n,p}, \varrho_{n,p}^{\Gamma_{n,p}})$ with $0 < \kappa < p$. Set $\lambda = \kappa - k$. Since we always have $C^{\kappa+1}(\Gamma_{\lambda,\kappa}, \varrho_{\lambda,\kappa}^{\Gamma_{\lambda,\kappa}}) = 0$,

$$\delta_\kappa : C^\kappa(\Gamma_{\lambda,\kappa}, \varrho_{\lambda,\kappa}^{\Gamma_{\lambda,\kappa}}) \to C^{\kappa+1}(\Gamma_{\lambda,\kappa}, \varrho_{\lambda,\kappa}^{\Gamma_{\lambda,\kappa}})$$

is the zero homomorphism, and hence $s_\kappa c \in C^\kappa(\Gamma_{\lambda,\kappa}, \varrho_{\lambda,\kappa}^{\Gamma_{\lambda,\kappa}})$ is a cocycle of the complex $\mathcal{C}(\Gamma_{\lambda,\kappa}, \varrho_{\lambda,\kappa}^{\Gamma_{\lambda,\kappa}})$, where

$$s_\kappa : C^\kappa(\Gamma_{n,p}, \varrho_{n,p}^{\Gamma_{n,p}}) \to C^\kappa(\Gamma_{\lambda,\kappa}, \varrho_{\lambda,\kappa}^{\Gamma_{\lambda,\kappa}})$$

is the homomorphism induced by the $(p-\kappa)$-th suspension. Therefore, for a $\Gamma_{\lambda,\kappa}^*$-map $f : M \to N$, the homology class $[s_\kappa c(f)] \in H_0^c(N; \mathbf{Z}_2)$ represented by $s_\kappa c(f)$ is well-defined. Note that its Poincaré dual in $H^\kappa(N; \mathbf{Z}_2)$ coincides with $\varphi_f([s_\kappa c])$, where $[s_\kappa c] \in H^\kappa(\Gamma_{\lambda,\kappa}, \varrho_{\lambda,\kappa}^{\Gamma_{\lambda,\kappa}})$ is the cohomology class represented by the cocycle $s_\kappa c$, and

$$\varphi_f : H^\kappa(\Gamma_{\lambda,\kappa}, \varrho_{\lambda,\kappa}^{\Gamma_{\lambda,\kappa}}) \to H^\kappa(N; \mathbf{Z}_2)$$

is the homomorphism induced by f. Furthermore, when the source manifold M is closed, $[s_\kappa c(f)]$ is well-defined as an element of $H_0(N; \mathbf{Z}_2)$.

Proposition 12.11. *Suppose that $\Gamma_{\lambda,\kappa}$ is transversely complete with respect to $\Gamma_{n,p}$, where $0 < \kappa < p$ and $p - n = \kappa - \lambda = k$. Then a cochain c in $C^\kappa(\Gamma_{n,p}, \varrho_{n,p}^{\Gamma_{n,p}})$ is a cocycle of the complex $\mathcal{C}(\Gamma_{n,p}, \varrho_{n,p}^{\Gamma_{n,p}})$ if and only if $[s_\kappa c(f)] = 0 \in H_0(N; \mathbf{Z}_2)$ (or equivalently, $\varphi_f([s_\kappa c]) = 0$ in $H^\kappa(N; \mathbf{Z}_2)$) for every $\Gamma_{\lambda,\kappa}^*$-map $f : M \to N$ such that both M and N are closed and that f is $\widetilde{\Gamma}_k$-cobordant to a nonsingular map.*

Proof. If c is a cocycle, then the cohomology class represented by $s_\kappa c$ lies in the image of

$$s_{\kappa *} : H^\kappa(\Gamma_{\lambda+1,\kappa+1}, \varrho_{\lambda+1,\kappa+1}^{\Gamma_{\lambda+1,\kappa+1}}) \to H^\kappa(\Gamma_{\lambda,\kappa}, \varrho_{\lambda,\kappa}^{\Gamma_{\lambda,\kappa}}).$$

Therefore, we have

$$[s_\kappa c(f)] = [s_\kappa c(f')] \in H_0(N; \mathbf{Z}_2) \tag{12.1}$$

for every f that is $\widetilde{\Gamma}_k$-cobordant to a nonsingular map f' by Lemma 12.4 (see also Remark 12.5). We see easily that (12.1) always vanishes, since $\kappa > 0$ and for a nonsingular map f', we have $s_\kappa c(f') = \emptyset$.

Conversely, suppose that $[s_\kappa c(f)] = 0 \in H_0(N; \mathbf{Z}_2)$ for every f as in the proposition. Let $\widetilde{\mathfrak{F}}$ be an arbitrary equivalence class of fibers in $\Gamma_{n,p}$ of codimension $\kappa + 1$ with respect to $\varrho_{n,p}^{\Gamma_{n,p}}$, and $g : M' \to N'$ be an element of $\mathcal{T}_{\mathrm{pr}}(n,p)$ such that the fiber of g over a point $y \in N'$ belongs to $\widetilde{\mathfrak{F}}$. By the proof of Lemma 7.3, we may assume that the stratum Σ containing y is of codimension $\kappa + 1$. Let N be the boundary of a sufficiently small $(\kappa+1)$-dimensional disk B in N' centered at y and transverse to Σ such that N is transverse to g and to all the strata of N'. Note that B corresponds to B_Σ in the argument just after Lemma 7.3. Then $f = g|_M : M \to N$ with $M = g^{-1}(N)$ is an element of $\mathcal{T}_{\mathrm{pr}}(\lambda, \kappa)$. Furthermore, since $\Gamma_{\lambda,\kappa}$ is transversely complete with respect to $\Gamma_{n,p}$ by our assumption, we see that f is a $\Gamma_{\lambda,\kappa}^*$-map.

It is easy to see that B contains a regular value y_0 of g with $y_0 \in B \setminus N$. Set $C = B - \mathrm{Int}\, B_0$, where B_0 is a closed disk neighborhood of y_0 in $B \setminus N$ consisting only of regular values of g and $\mathrm{Int}\, B_0$ denotes its interior as a subspace of B. Note that C is diffeomorphic to $S^\kappa \times [0,1]$. Then, we see that $g|_{\widetilde{C}} : \widetilde{C} \to C$ with $\widetilde{C} = g^{-1}(C)$ gives a $\widetilde{\Gamma}_k$-cobordism between f and a nonsingular map. Hence, by our assumption, $s_\kappa c(f)$ consists of an even number of points. This means that the coefficient of $\widetilde{\mathfrak{F}}$ in $\delta_\kappa(s_\kappa c)$ is zero (see (8.2)). Since this holds for an arbitrary $\widetilde{\mathfrak{F}}$ of codimension $\kappa+1$, we have $\delta_\kappa(s_\kappa c) = 0$. This completes the proof. □

Now let us apply the above proposition to a specific but important situation as follows.

Corollary 12.12. *Let us consider the complex*

$$\mathcal{C}(\mathcal{S}^0_{\text{pr}}(5,4)^{\text{ori}}, \varrho^0_{5,4}(2)) \tag{12.2}$$

of singular fibers for proper C^0 stable maps of orientable 5-dimensional manifolds into 4-dimensional manifolds with respect to the C^0 equivalence modulo two circle components. Let $\widehat{\text{III}}^8_{\text{o}}$ (or $\widehat{\text{III}}^8_{\text{e}}$) be the C^0 equivalence class modulo two circle components of the suspension of III^8_{o} (resp. III^8_{e}). Then $\widehat{\text{III}}^8_{\text{o}} + \widehat{\text{III}}^8_{\text{e}}$ is a 3-cocycle of the complex (12.2) *and represents a nontrivial cohomology class in $H^3(\mathcal{S}^0_{\text{pr}}(5,4)^{\text{ori}}, \varrho^0_{5,4}(2))$.*

For notations, refer to Fig. 3.4 and Proposition 9.4.

Proof. As in Proposition 3.1, we can obtain a similar characterization of proper C^∞ stable maps of 5-dimensional manifolds into 4-dimensional manifolds (for details, see [50]). Using this and Proposition 3.1 itself, we can show that $\mathcal{S}^0_{\text{pr}}(4,3)^{\text{ori}}$ is transversely complete with respect to $\mathcal{S}^0_{\text{pr}}(5,4)^{\text{ori}}$. Furthermore, an argument similar to that of the proof of Corollary 3.9 shows that two elements of $\mathcal{S}^0_{\text{pr}}(4,3)^{\text{ori}*}$ are C^0 equivalent modulo two circle components if and only if so are their suspensions in $\mathcal{S}^0_{\text{pr}}(5,4)^{\text{ori}*}$. Hence, we have $s_3(\widehat{\text{III}}^8_{\text{o}} + \widehat{\text{III}}^8_{\text{e}}) = \text{III}^8_{\text{o}} + \text{III}^8_{\text{e}}$.

Now suppose that a C^0 stable map $f : M \to N$ of a closed orientable 4-manifold into a closed 3-manifold is $\widetilde{\mathcal{S}}^0_{\text{pr}}(-1)$-cobordant to a nonsingular map. Since the source manifold of a nonsingular map always has zero Euler characteristic, we see that the Euler characteristic of M should be even. Hence, by Theorem 5.1, the number of elements in the set $(\text{III}^8_{\text{o}} + \text{III}^8_{\text{e}})(f)$ should be even, and hence it represents the trivial homology class in $H_0(N; \mathbf{Z}_2)$. Then, by Proposition 12.11, we see that $\widehat{\text{III}}^8_{\text{o}} + \widehat{\text{III}}^8_{\text{e}}$ is a cocycle of the complex (12.2).

Note that there does exist a closed orientable 4-manifold whose Euler characteristic is odd. Let $g : M' \to N'$ be a C^0 stable map of such a 4-manifold M' into a 3-manifold N'. Then, again by Theorem 5.1, we see that, for the homomorphism

$$\varphi_g : H^3(\mathcal{S}^0_{\text{pr}}(4,3)^{\text{ori}}, \varrho^0_{4,3}(2)) \to H^3_c(N'; \mathbf{Z}_2)$$

induced by g, we have $\varphi_g([\text{III}^8_{\text{o}} + \text{III}^8_{\text{e}}]) \neq 0$. This shows that the cohomology class $s_{3*}([\widehat{\text{III}}^8_{\text{o}} + \widehat{\text{III}}^8_{\text{e}}]) = [\text{III}^8_{\text{o}} + \text{III}^8_{\text{e}}]$ does not vanish, and hence that the cohomology class $[\widehat{\text{III}}^8_{\text{o}} + \widehat{\text{III}}^8_{\text{e}}]$ is nontrivial. This completes the proof. □

The above corollary justifies the prediction given in Remark 10.12.

Let us end this chapter by the following proposition concerning Problem 11.14.

Proposition 12.13. *Suppose that $\Gamma_{\lambda,\kappa}$ is transversely complete with respect to $\Gamma_{n,p}$, where $0 < \kappa < p$ and $p - n = \kappa - \lambda = k$. Then the following two are equivalent.*

(1) *A cochain $c \in C^{\kappa}(\Gamma_{n,p}, \varrho_{n,p}^{\Gamma_{n,p}})$ is a coboundary if and only if $[s_{\kappa}c(f)] = 0 \in H_0(N; \mathbf{Z}_2)$ (or equivalently, $\varphi_f([s_{\kappa}c]) = 0$ in $H^{\kappa}(N; \mathbf{Z}_2)$) for every $\Gamma_{\lambda,\kappa}^*$-map $f : M \to N$ such that both M and N are closed.*

(2) *If $\alpha \in H^{\kappa}(\Gamma_{n,p}, \varrho_{n,p}^{\Gamma_{n,p}})$ is nonzero, then there exists a $\Gamma_{n,p}^*$-map $g : M' \to N'$ such that $\varphi_g(\alpha) \neq 0$ in $H^{\kappa}(N'; \mathbf{Z}_2)$.*

Proof. (1) $\Longrightarrow$ (2). Suppose that $\varphi_g(\alpha) = 0$ in $H^{\kappa}(N'; \mathbf{Z}_2)$ for all $\Gamma_{n,p}^*$-map $g : M' \to N'$. Then, by Lemma 11.11, for every $\Gamma_{\lambda,\kappa}^*$-map $f : M \to N$, we have $\varphi_f(s_{\kappa *}\alpha) = 0$ in $H^{\kappa}(N; \mathbf{Z}_2)$. Now item (1) implies that $\alpha = 0$. This is a contradiction.

(2) $\Longrightarrow$ (1). Suppose that c is a coboundary. Then $\varphi_f([s_{\kappa}c]) = 0$ in $H^{\kappa}(N; \mathbf{Z}_2)$, since $[s_{\kappa}c] = s_{\kappa *}[c] = 0$. Conversely, suppose that $[s_{\kappa}c(f)] = 0 \in H_0(N; \mathbf{Z}_2)$ for every $\Gamma_{\lambda,\kappa}^*$-map $f : M \to N$ such that both M and N are closed. By Proposition 12.11, c is a cocycle. Then, by Lemma 11.10, $\varphi_g([c]) = 0$ for every $\Gamma_{n,p}^*$-map g. Then item (2) implies that $[c] = 0$; i.e. c is a coboundary. This completes the proof. □

Note that all the results in this section hold also for the universal complexes of co-orientable singular fibers and the cohomology groups with $\mathbf{Z}$-coefficients, provided that the target manifolds are oriented, except for Corollary 12.12.

Remark 12.14. Recall that $\mathcal{T}_{\mathrm{pr}}(\lambda, \kappa)$ is always transversely complete with respect to $\mathcal{T}_{\mathrm{pr}}(n, p)$. Thus, in view of Proposition 12.13, a special case of Problem 11.14 can be interpreted as follows at least for Thom maps. A cochain $c \in C^{\kappa}(\mathcal{T}_{\mathrm{pr}}(n, p), \varrho_{n,p}^0)$ with $0 < \kappa < p$ of the universal complex is a cocycle if and only if $s_{\kappa}c(f)$ is null-homologous for all f cobordant to a nonsingular map. Is it true that a cocycle c is a coboundary if and only if $s_{\kappa}c(f)$ is null-homologous for all f?

13

Cobordism of Maps with Prescribed Local Singularities

In this chapter, we consider another cobordism relation which is slightly different from the one given in the previous chapter.

Let us consider a C^∞ stable (mono-)germ $\eta : (\mathbf{R}^n, 0) \to (\mathbf{R}^{n+k}, 0)$ of codimension k. We define its *suspension* $\Sigma\eta : (\mathbf{R}^{n+1}, 0) \to (\mathbf{R}^{n+1+k}, 0)$ by $\Sigma\eta(u, t) = (\eta(u), t)$ for $u \in \mathbf{R}^n$ and $t \in \mathbf{R}$. For a fixed $k \in \mathbf{Z}$, let us consider the set of C^∞ stable map germs of codimension k, and the equivalence relation generated by the C^∞ right-left equivalence and the suspension. We call such an equivalence class a *singularity type* (see [40]).

There is a hierarchy of singularity types. A singularity type A is said to be *under* another singularity type B if for a representative $f : (\mathbf{R}^n, 0) \to (\mathbf{R}^{n+k}, 0)$ of A, there is a germ of B arbitrary close to f, in the sense that there are points x arbitrary close to the origin of $\mathbf{R}^n$ such that the germ of f at x belongs to B. In this case, we also say that B is *over* A. (Compare this with Definition 8.18.)

Let τ be an ascending set of singularity types.

Definition 13.1. We say that a smooth map $f : M \to N$ between smooth manifolds is a *τ-map* if its singularities (as mono-germs) in the source manifold M all lie in τ. If M is a manifold with boundary, then we also suppose that $f(\partial M) \subset \partial N$ and for collar neighborhoods C and C' of ∂M and ∂N respectively, we have $f|_C = \Sigma(f|_{\partial M})$.

Definition 13.2. For a smooth manifold N, two τ-maps $f_0 : M_0 \to N$ and $f_1 : M_1 \to N$ of closed manifolds M_0 and M_1 are said to be *τ-cobordant* if there exist a compact manifold W with boundary the disjoint union $M_0 \amalg M_1$, and a τ-map $F : W \to N \times [0, 1]$ such that $f_i = F|_{M_i} : M_i \to N \times \{i\}$, $i = 0, 1$. We call F a *τ-cobordism* between f_0 and f_1.

When M_i are oriented and W can be taken to be oriented so that $\partial W = (-M_0) \amalg M_1$, then we say that f_0 and f_1 are *oriented τ-cobordant.*

Lemma 13.3. *Every τ-map of a closed manifold is τ-cobordant to a τ-map which is a Thom map.*

Proof. Suppose that a τ-map $f : M \to N$ is given, where M is a closed manifold. Then there exists a τ-map $\widetilde{f} : M \to N$ which is a Thom map and which is sufficiently close to f in the mapping space $C^\infty(M, N)$, since the set of Thom maps is dense in $C^\infty(M, N)$ and the local singularities of f are all C^∞ stable (and hence the set of all τ-maps is open in the mapping space). In particular, we may assume that f and $\widetilde{f}$ are homotopic through τ-maps and hence are τ-cobordant. This completes the proof. □

Remark 13.4. Suppose that two τ-maps $f_i : M_i \to N$, $i = 0, 1$, of closed manifolds are Thom maps. If they are τ-cobordant, then a τ-cobordism between them can be chosen as a Thom map. This is proved by first taking any τ-cobordism and then by approximating it by a Thom map.

In what follows, we fix the codimension $k \in \mathbf{Z}$. For an ascending set τ of singularity types of codimension k and for a dimension pair (n, p) with $p - n = k$, let us denote by $\tau(n, p)$ the set of all proper Thom maps which are τ-maps. Furthermore, we set

$$\widetilde{\tau}(k) = \bigcup_p \tau(p - k, p),$$

and let us consider a stable system of admissible equivalence relations $\mathcal{R}_k^\tau = \{\varrho_{p-k,p}^\tau\}_p$ for the fibers of elements of $\widetilde{\tau}(k)$. Note that the set $\widetilde{\tau}(k)$ is closed under suspension.

Definition 13.5. Let $f : M \to N$ be an arbitrary τ-map, which may not necessarily be a Thom map, where we assume that M is closed. Then by Lemma 13.3, f is τ-cobordant to a τ-map $\widetilde{f} : M \to N$ which is a Thom map. Then we define

$$\varphi_f : \operatorname{Im} s_{\kappa *} \to H^\kappa(N; \mathbf{Z}_2)$$

by $\varphi_f = \varphi_{\widetilde{f}}|_{\operatorname{Im} s_{\kappa *}}$, where

$$s_{\kappa *} : H^\kappa(\tau(n+1, p+1), \varrho_{n+1,p+1}^\tau) \to H^\kappa(\tau(n, p), \varrho_{n,p}^\tau)$$

is the homomorphism induced by the suspension, and

$$\varphi_{\widetilde{f}} : H^\kappa(\tau(n, p), \varrho_{n,p}^\tau) \to H^\kappa(N; \mathbf{Z}_2)$$

is the homomorphism induced by the Thom map $\widetilde{f}$. The homomorphism φ_f is well-defined by virtue of Lemma 12.4 together with Remark 13.4.

By Lemma 12.4, we see that if f_0 and f_1 are τ-maps of closed manifolds into a p-dimensional manifold N which are τ-cobordant, then $\varphi_{f_0} = \varphi_{f_1}$. In other words, the correspondence $f \mapsto \varphi_f$ defines a τ-cobordism invariant of τ-maps into N.

Remark 13.6. If τ is big enough, or more precisely, if the space of τ-maps is always dense in the corresponding mapping space, then for every smooth map $f : M \to N$ of a closed manifold, we can define φ_f to be $\varphi_{\widetilde{f}}$, where $\widetilde{f}$ is an approximation of f which is a τ-map. Then, we can show that this is well-defined, and that it defines a *bordism invariant* of smooth maps into N, where two smooth maps $f_0 : M_0 \to N$ and $f_1 : M_1 \to N$ of closed manifolds M_0 and M_1 are said to be *bordant* if there exist a compact manifold W with boundary the disjoint union $M_0 \amalg M_1$, and a smooth map $F : W \to N \times [0,1]$ such that $f_i = F|_{M_i} : M_i \to N \times \{i\}$, $i = 0, 1$ (for details, see [8]). In particular, if N is contractible, it defines a cobordism invariant of the source manifold.

Remark 13.7. So far, we have considered Thom maps which are τ-maps. It is easy to see that we could as well consider C^0 stable maps which are Thom maps (or C^∞ stable maps for nice dimension pairs (n,p) in the sense of Mather [32]) instead of Thom maps, since the corresponding sets are dense in the mapping spaces. Let us denote by $\tau^0(n,p)$ the set of C^0 stable maps in $\mathcal{T}_{\mathrm{pr}}(n,p)$ which are τ-maps. Then, for a τ-map $f : M \to N$ with M being closed, we can define the homomorphism

$$\varphi_f : \mathrm{Im}\, s^0_{\kappa *} \to H^\kappa(N; \mathbf{Z}_2),$$

which is a τ-cobordism invariant, where

$$s^0_{\kappa *} : H^\kappa(\tau^0(n+1,p+1), \varrho^\tau_{n+1,p+1}) \to H^\kappa(\tau^0(n,p), \varrho^\tau_{n,p})$$

is the homomorphism induced by the suspension.

In fact, we can show that the diagram

$$\begin{array}{ccc} H^\kappa(\tau(n+1,p+1), \varrho^\tau_{n+1,p+1}) & \xrightarrow{s_{\kappa *}} & H^\kappa(\tau(n,p), \varrho^\tau_{n,p}) \\ {\scriptstyle \pi_{\tau(n+1,p+1),\tau^0(n+1,p+1)*}} \downarrow & & \downarrow {\scriptstyle \pi_{\tau(n,p),\tau^0(n,p)*}} \\ H^\kappa(\tau^0(n+1,p+1), \varrho^\tau_{n+1,p+1}) & \xrightarrow{s^0_{\kappa *}} & H^\kappa(\tau^0(n,p), \varrho^\tau_{n,p}) \end{array}$$

is commutative, and that

$$\varphi_f : \mathrm{Im}\, s_{\kappa *} \to H^\kappa(N; \mathbf{Z}_2)$$

coincides with the composition of the natural homomorphism induced by the projection

$$\pi_{\tau(n,p),\tau^0(n,p)*}|_{\mathrm{Im}\, s_{\kappa *}} : \mathrm{Im}\, s_{\kappa *} \to \mathrm{Im}\, s^0_{\kappa *}$$

and

$$\varphi_f : \mathrm{Im}\, s^0_{\kappa *} \to H^\kappa(N; \mathbf{Z}_2).$$

Let us consider τ-maps into $N = N' \times \mathbf{R}$, where N' is a $(p-1)$-dimensional manifold. Then, the set of all τ-cobordism classes of τ-maps of closed manifolds into N, denoted by $\mathrm{Cob}_\tau(N)$, forms an abelian group with respect to the "far away disjoint union". (When we take the orientations into account, we denote the corresponding abelian group by $\mathrm{Cob}_\tau^{\mathrm{ori}}(N)$.) More precisely, for two τ-maps $f_i : M_i \to N$ of closed manifolds M_i, $i = 0, 1$, there exists a real number r such that $f_0(M_0) \cap (T_r \circ f_1(M_1)) = \emptyset$, where the diffeomorphism $T_r : N' \times \mathbf{R} \to N' \times \mathbf{R}$ is defined by $T_r(x,t) = (x, t+r)$. Then, it is not difficult to show that the τ-cobordism class of the disjoint union of the two maps f_0 and $T_r \circ f_1$ depends only on the τ-cobordism classes of f_0 and f_1. Furthermore, the resulting τ-cobordism class does not change even if we interchange f_0 and f_1. Thus, we define $[f_0] + [f_1] = [f_0 \amalg (T_r \circ f_1)]$, where $[*]$ denotes the τ-cobordism class of $*$. The neutral element is the map of the empty set, and the inverse element of a τ-map $f : M \to N' \times \mathbf{R}$ is given by $-f : M \to N' \times \mathbf{R}$ defined by $-f = R \circ f$, where $R : N' \times \mathbf{R} \to N' \times \mathbf{R}$ is the diffeomorphism defined by $R(x,t) = (x, -t)$. (When we take the orientations into account, the source manifold of $-f$ is understood to be $-M$.)

Then, the following is a direct consequence of the above definitions.

Proposition 13.8. *In the above situation, the map*

$$\Phi_\kappa : \mathrm{Cob}_\tau(N) \to \mathrm{Hom}\,(\mathrm{Im}\, s_{\kappa *}, H^\kappa(N; \mathbf{Z}_2))$$

defined by $\Phi_\kappa([f]) = \varphi_f$ for a τ-maps f of a closed manifold into N is a homomorphism of abelian groups for every κ, where

$$s_{\kappa *} : H^\kappa(\tau(n+1, p+1), \varrho^\tau_{n+1,p+1}) \to H^\kappa(\tau(n,p), \varrho^\tau_{n,p})$$

is the homomorphism induced by the suspension.

Note that a similar map

$$\Phi_\kappa : \mathrm{Cob}_\tau(N) \to \mathrm{Hom}\,(\mathrm{Im}\, s^0_{\kappa *}, H^\kappa(N; \mathbf{Z}_2))$$

can also be defined and is a homomorphism of abelian groups for every κ (see Remark 13.7).

We do not know if the homomorphism $\oplus_\kappa \Phi_\kappa$ is injective or not for some $\varrho^\tau_{n,p}$ and $\varrho^\tau_{n+1,p+1}$.

Remark 13.9. Note that the above proposition holds also for $\widetilde{\Gamma}_k$-maps in the sense of §8.6 or Chap. 12. However, we do not know if the group operation defined on the set of cobordism classes is commutative or not. If $N = N'' \times \mathbf{R}^2$ for some $(p-2)$-dimensional manifold N'', then we can show that the resulting group is abelian.

Note that all the results in this chapter hold also for the universal complexes of co-orientable singular fibers and the cohomology groups with $\mathbf{Z}$-coefficients, provided that the target manifolds are oriented.

14

Examples of Cobordism Invariants

In this chapter, we shall construct explicit cobordism invariants in specific situations following the procedure introduced in the previous chapters. Throughout the chapter, the codimension will always be equal to -1. Furthermore, we shall work only with nice dimension pairs, and we shall consider C^0 stable maps instead of Thom maps following Remark 13.7.

14.1 Cobordism of Stable Maps

Let τ be the set of singularity types corresponding to a regular point and a Morin singularity [34], i.e., a fold point, a cusp point, a swallowtail, etc. Note that if the dimension of the source manifold is less than or equal to 4, this set is big enough in the sense of Remark 13.6.

Let us consider C^0 stable maps of surfaces and 3-manifolds. By Proposition 9.7, the first cohomology group of the universal complex of singular fibers

$$\mathcal{C}(\mathcal{S}^0_{\mathrm{pr}}(3,2), \varrho^0_{3,2}(2)) = \mathcal{C}(\tau^0(3,2), \varrho^0_{3,2}(2))$$

with respect to the C^0 equivalence modulo two circle components for $\tau^0(3,2)$-maps is isomorphic to $\mathbf{Z}_2 \oplus \mathbf{Z}_2$ and is generated by $\alpha_1 = [\widetilde{\mathrm{I}}^0_{\mathrm{o}} + \widetilde{\mathrm{I}}^1_{\mathrm{e}}] = [\widetilde{\mathrm{I}}^0_{\mathrm{e}} + \widetilde{\mathrm{I}}^1_{\mathrm{o}}]$ and $\alpha_2 = [\widetilde{\mathrm{I}}^2_{\mathrm{o}} + \widetilde{\mathrm{I}}^2_{\mathrm{e}}]$. In the following, let

$$s^0_{1*} : H^1(\tau^0(3,2), \varrho^0_{3,2}(2)) \to H^1(\tau^0(2,1), \varrho^0_{2,1}(2))$$

be the homomorphism induced by the suspension.

Let us first consider $\alpha_2 = [\widetilde{\mathrm{I}}^2_{\mathrm{o}} + \widetilde{\mathrm{I}}^2_{\mathrm{e}}]$. For a C^∞ stable map $f : M \to N$ of a closed surface into a connected 1-dimensional manifold N, $(s^0_{1*}\alpha_2)(f) \in H_0(N; \mathbf{Z}_2) \cong \mathbf{Z}_2$ is nothing but the number modulo two of the singular fibers as depicted in Fig. 2.2 (3). By Lemma 12.4 and Remark 13.6, this is a bordism invariant. On the other hand, by Corollary 2.4 and Remark 2.9, the number modulo two coincides with the parity of the Euler characteristic of the source surface M. Thus, $(s^0_{1*}\alpha_2)(f)$ coincides with the parity of the Euler characteristic of its source surface. When $N = \mathbf{R}$, this is a complete bordism (or τ-cobordism) invariant for (τ-)maps of closed surfaces into N.

For $\alpha_1 = [\widetilde{\mathrm{I}}^0_{\mathrm{o}} + \widetilde{\mathrm{I}}^1_{\mathrm{e}}] = [\widetilde{\mathrm{I}}^0_{\mathrm{e}} + \widetilde{\mathrm{I}}^1_{\mathrm{o}}]$, we have the following.

Lemma 14.1. *For every $\tau^0(2,1)$-map $f : M \to N$ of a closed surface M into a 1-dimensional manifold N, $(s^0_{1*}\alpha_1)(f) \in H_0(N; \mathbf{Z}_2)$ vanishes.*

Proof. Let $\mathbf{O}_{\mathrm{o}}(f)$ be the set

$$\{y \in N \,:\, y \text{ is a regular value of } f \text{ and } b_0(f^{-1}(y)) \text{ is odd}\}.$$

Since every 1-dimensional manifold is orientable, we give an orientation to N. Then each connected component of $\overline{\mathbf{O}_{\mathrm{o}}(f)}$, which is either an arc or a circle, has an induced orientation. Note that the end points of the arc components of $\overline{\mathbf{O}_{\mathrm{o}}(f)}$ correspond to

$$(\widetilde{\mathrm{I}}^0_{\mathrm{o}} + \widetilde{\mathrm{I}}^1_{\mathrm{e}} + \widetilde{\mathrm{I}}^0_{\mathrm{e}} + \widetilde{\mathrm{I}}^1_{\mathrm{o}})(f).$$

For an arc component, we say that it is of type $++$ (or $--$) if the number of connected components of a regular fiber of f increases (resp. decreases) by one when the target point passes through its starting point and also when it passes through its terminal point. We say that it is of type $+-$ (or $-+$) if the number increases (resp. decreases) by one when the target point passes through its starting point and it decreases (resp. increases) by one when it passes through its terminal point. In this way, the arc components of $\overline{\mathbf{O}_{\mathrm{o}}(f)}$ can be classified into these four types. We denote by $n(++)$, $n(--)$, $n(+-)$ and $n(-+)$ the numbers of arc components of types $++$, $--$, $+-$ and $-+$ respectively.

Then, it is easy to show that

$$|\widetilde{\mathrm{I}}^0_{\mathrm{o}}(f)| + |\widetilde{\mathrm{I}}^1_{\mathrm{e}}(f)| = n(++) + n(--) + 2n(+-),$$
$$|\widetilde{\mathrm{I}}^0_{\mathrm{e}}(f)| + |\widetilde{\mathrm{I}}^1_{\mathrm{o}}(f)| = n(++) + n(--) + 2n(-+).$$

Since we should have $n(++) = n(--)$, we obtain

$$|\widetilde{\mathrm{I}}^0_{\mathrm{o}}(f)| + |\widetilde{\mathrm{I}}^1_{\mathrm{e}}(f)| \equiv |\widetilde{\mathrm{I}}^0_{\mathrm{e}}(f)| + |\widetilde{\mathrm{I}}^1_{\mathrm{o}}(f)| \equiv 0 \pmod 2.$$

This implies that $(s^0_{1*}\alpha_1)(f) = 0$ in $H_0(N; \mathbf{Z}_2)$. This completes the proof. □

Remark 14.2. Note that $s^0_{1*}\alpha_1$ does not vanish as an element of the cohomology group $H^1(\tau^0(2,1), \varrho^0_{2,1}(2))$. Hence, the above lemma shows that even if we take a nontrivial cohomology class of the universal complex with respect to an admissible equivalence relation, the corresponding homology class in the target manifold can be trivial. Hence, the answer to the problem mentioned in Problem 11.14 is negative in general, if we replace the C^0 equivalence relation $\varrho^0_{n,p}$ with an arbitrary admissible equivalence relation, at least for the cohomology group (11.2). See also Remark 12.14.

Let us consider the homomorphism

$$\varepsilon_{\varrho^0_{2,1}(2),\varrho^0_{2,1}*} : H^1(\mathcal{S}^0_{\mathrm{pr}}(2,1), \varrho^0_{2,1}(2)) \to H^1(\mathcal{S}^0_{\mathrm{pr}}(2,1), \varrho^0_{2,1})$$

induced by the cochain map

$$\varepsilon_{\varrho^0_{2,1}(2),\varrho^0_{2,1}} : \mathcal{C}(\mathcal{S}^0_{\mathrm{pr}}(2,1), \varrho^0_{2,1}(2)) \to \mathcal{C}(\mathcal{S}^0_{\mathrm{pr}}(2,1), \varrho^0_{2,1})$$

defined in §8.6. If the image of $s^0_{1*}\alpha_1 \in H^1(\mathcal{S}^0_{\mathrm{pr}}(2,1), \varrho^0_{2,1}(2))$ by $\varepsilon_{\varrho^0_{2,1}(2),\varrho^0_{2,1}*}$ is nontrivial, then the problem mentioned in Problem 11.14 is negatively solved. The author conjectures that $\varepsilon_{\varrho^0_{2,1}(2),\varrho^0_{2,1}*}(s^0_{1*}\alpha_1) \neq 0$.

In [63, 64], Yamamoto considers an equivalence relation among the fibers of a given map which takes into account their positions from a global viewpoint. In other words, even if two fibers are C^0 equivalent, if their positions are different from each other in a certain global sense, then one considers them to be nonequivalent. Probably, we can construct universal complexes of singular fibers with respect to such "global" equivalence relations. Then, the author conjectures that for such a universal complex with respect to a certain global equivalence relation, the answer to the problem mentioned in Problem 11.14 should be positive.

If we consider a C^0 stable map $f : M \to N$ of a closed 3-manifold into a surface, then $\alpha_1(f)$ and $\alpha_2(f)$ are defined as elements of $H_1(N; \mathbf{Z}_2)$. We see that $\alpha_2(f)$ can be nontrivial by the example constructed as follows.

Let $g : \mathbf{R}P^2 \to \mathbf{R}$ be an arbitrary Morse function. Note that $(s^0_{1*}\alpha_2)(g)$ is nontrivial by Corollary 2.4. We define $f = g \times \mathrm{id}_{S^1} : \mathbf{R}P^2 \times S^1 \to \mathbf{R} \times S^1$. Then we see that $\alpha_2(f)$ does not vanish in $H_1(\mathbf{R} \times S^1; \mathbf{Z}_2)$. (This implies, for example, that $f : \mathbf{R}P^2 \times S^1 \to \mathbf{R} \times S^1$ is not bordant to a constant map.)

On the other hand, $\alpha_1(f)$ always vanishes. This follows from Lemma 11.10, since $(s^0_{1*}\alpha_1)(h)$ always vanishes for a $\tau^0(2,1)$-map h of a closed surface into a 1-dimensional manifold as mentioned above. Note that $\tau^0(2,1)$ is transversely complete with respect to $\tau^0(3,2)$.

14.2 Cobordism of Fold Maps

Let us now consider an example of τ which is not big in the sense of Remark 13.6. Let τ be the set of singularity types corresponding to a regular point and a fold point. In this case, a τ-map is called a *fold map.* (Recall that this notion was already introduced in Chap. 6). In the following, we denote by $\tau^0(n,p)^{\mathrm{ori}}$ the set of all C^0 equivalence classes of fibers for proper C^0 stable τ-maps in $\mathcal{T}_{\mathrm{pr}}(n,p)$ of orientable n-dimensional manifolds.

Then the following proposition can be proved. Details are left to the reader.

Proposition 14.3. *The cohomology groups of the universal complex*

$$\mathcal{CO}(\tau^0(3,2)^{\mathrm{ori}}, \varrho^0_{3,2}(2))$$

of co-orientable singular fibers for proper C^0 stable fold maps of orientable 3-manifolds into surfaces with respect to the C^0 equivalence modulo two circle components are given as follows:

$$H^0(\mathcal{CO}(\tau^0(3,2)^{\mathrm{ori}}, \varrho^0_{3,2}(2))) \cong \mathbf{Z} \ (\textit{generated by } [\mathbf{0}_{\mathrm{o}} + \mathbf{0}_{\mathrm{e}}]),$$

$$H^1(\mathcal{CO}(\tau^0(3,2)^{\mathrm{ori}}, \varrho^0_{3,2}(2))) \cong \mathbf{Z} \oplus \mathbf{Z} \ (\textit{generated by } \alpha_1 = [\widetilde{\mathrm{I}}^0_{\mathrm{o}} + \widetilde{\mathrm{I}}^1_{\mathrm{e}}] = [\widetilde{\mathrm{I}}^0_{\mathrm{e}} + \widetilde{\mathrm{I}}^1_{\mathrm{o}}],$$
$$\alpha_2 = [\widetilde{\mathrm{I}}^0_{\mathrm{o}} + \widetilde{\mathrm{I}}^0_{\mathrm{e}}], \textit{ and } \alpha_3 = [\widetilde{\mathrm{I}}^1_{\mathrm{o}} + \widetilde{\mathrm{I}}^1_{\mathrm{e}}]$$
$$\textit{with } 2\alpha_1 = \alpha_2 + \alpha_3),$$

where for a C^0 equivalence class $\mathfrak{F}$ of fibers, $\mathfrak{F}_{\mathrm{o}}$ (or $\mathfrak{F}_{\mathrm{e}}$) denotes the C^0 equivalence class modulo two circle components represented by $\mathfrak{F}_\ell$ with ℓ odd (resp. even), and $[*]$ denotes the cohomology class represented by the cocycle $*$.

Let $f : M \to \mathbf{R}$ be a Morse function, which is a fold map, of a closed oriented surface M. Then $(s^0_{1*}\alpha_2)(f) \in H_0(\mathbf{R}; \mathbf{Z}) \cong \mathbf{Z}$ coincides with $\max(f) - \min(f)$, where $\max(f)$ (or $\min(f)$) is the number of local maxima (resp. minima) of the Morse function f. Furthermore, $(s^0_{1*}\alpha_3)(f)$ coincides with the τ-cobordism invariant introduced in [22]. Since we can show that $(s^0_{1*}\alpha_1)(f)$ always vanishes as in Lemma 14.1, we have $(s^0_{1*}\alpha_2)(f) = -(s^0_{1*}\alpha_3)(f)$.

Note that by [22], two Morse functions f_0 and f_1 on closed oriented surfaces are oriented τ-cobordant if and only if $(s^0_{1*}\alpha_2)(f_0) = (s^0_{1*}\alpha_2)(f_1)$. In other words, the cohomology class $s^0_{1*}\alpha_2$ of the universal complex of co-orientable singular fibers with respect to the C^0 equivalence modulo two circle components gives a complete invariant for τ-cobordisms of τ-maps of oriented surfaces into $\mathbf{R}$.

Part III

Epilogue

15

Applications

In this chapter, we give some applications of the ideas developed in Chap. 7 to the topology of generic maps.

First, we prepare some lemmas.

Lemma 15.1. *Let W be a compact m-dimensional manifold such that its boundary is a disjoint union of open and closed subsets V_0 and V_1. If there exists a Morse function $g : W \to \mathbf{R}$ such that $g(W) = [a, b]$ for some $a < b$, $V_0 = g^{-1}(a)$, $V_1 = g^{-1}(b)$, and that g has a unique critical point in the interior of W, then the difference between the Euler characteristics of V_0 and V_1 is equal to ± 2, provided that m is odd.*

Proof. Let λ be the index of the critical point. Then by Morse theory, we see that V_1 is diffeomorphic to

$$(V_0 \setminus \mathrm{Int}(S^{\lambda-1} \times D^{m-\lambda})) \cup (D^\lambda \times S^{m-\lambda-1}).$$

Then the result follows immediately. □

Definition 15.2. Let V_0 and V_1 be closed oriented $(4k+1)$-dimensional manifolds with $k \geq 0$. Suppose that there exists an oriented cobordism W between V_0 and V_1. Then, we define $d(V_0, V_1)$ to be the Euler characteristic modulo two of W. Since every closed orientable $(4k+2)$-dimensional manifold has even Euler characteristic, $d(V_0, V_1) \in \mathbf{Z}_2$ does not depend of the choice of W. In fact, $d(V_0, V_1)$ coincides with the difference between the semi-characteristics $\chi^*(V_0)$ and $\chi^*(V_1)$ with respect to any coefficient field (see [31] and [21, §5]).

Then the following lemma follows from the very definition.

Lemma 15.3. *Let W be a compact $(4k+2)$-dimensional oriented manifold such that its boundary is a disjoint union of open and closed subsets V_0 and V_1. If there exists a Morse function $g : W \to \mathbf{R}$ such that $g(W) = [a, b]$ for some $a < b$, $V_0 = g^{-1}(a)$, $V_1 = g^{-1}(b)$, and that g has a unique critical point in the interior of W, then $d(V_0, V_1)$ defined above is equal to $1 \in \mathbf{Z}_2$.*

With the help of the above lemmas, we prove the following. Recall that a smooth map between smooth manifolds is a *Boardman map* if its jet extensions are transverse to all the Thom-Boardman subbundles (see [5] and [16, Chapter VI, §5]). Furthermore, such a map satisfies the *normal crossing condition* if its restrictions to the Thom-Boardman strata intersect in general position (for more details, see [16, Chapter VI, §5]).

Proposition 15.4. *Let $f : M \to N$ be a Boardman map of a closed n-dimensional manifold M into a p-dimensional manifold with $n \geq p$. Suppose either that $n - p$ is even, or that $n - p \equiv 1 \pmod 4$ and M is orientable. Then $f_*[S(f)]_2 = 0 \in H_{p-1}(N; \mathbf{Z}_2)$, where $[S(f)]_2 \in H_{p-1}(M; \mathbf{Z}_2)$ is the $\mathbf{Z}_2$-homology class represented by the singular set $S(f)$ of f.*

Proof. We may assume that N is connected. We may also assume that f satisfies the normal crossing condition by perturbing f slightly. When $n-p \equiv 1 \pmod 4$, we fix an orientation of M. Take a regular value $y_0 \in N$ of f and fix it, where we take $y_0 \in N \smallsetminus f(M)$ if N is open. Let R be the closure of the set of points $y \in N \smallsetminus f(S(f))$ such that

$$\frac{\chi(f^{-1}(y)) - \chi(f^{-1}(y_0))}{2}$$

is odd for $n - p \equiv 0 \pmod 2$ and that

$$d(f^{-1}(y_0), f^{-1}(y_1)) \equiv 1 \pmod 2$$

for $n-p \equiv 1 \pmod 4$. (Note that in the latter case, both $f^{-1}(y_0)$ and $f^{-1}(y_1)$ are orientable manifolds.) Note that if A is an embedded arc connecting y and y_0 transverse to f, then $f^{-1}(A)$ gives a (oriented) cobordism between $f^{-1}(y)$ and $f^{-1}(y_0)$, and hence $\chi(f^{-1}(y)) - \chi(f^{-1}(y_0))$ is always an even integer for $n - p \equiv 0 \pmod 2$ and $d(f^{-1}(y_0), f^{-1}(y_1)) \in \mathbf{Z}_2$ is well-defined for $n - p \equiv 1 \pmod 4$. Then it is easy to see that R is compact.

Since f is a Boardman map, $S(f)$ is naturally stratified into the Thom-Boardman strata, and the top dimensional strata of $S(f)$ consist of fold points. Let J be an arc embedded in N such that J intersects $f(S(f))$ transversely at a unique interior point z such that $f^{-1}(z) \cap S(f)$ consists of a fold point. Then by applying Lemmas 15.1 and 15.3 to the (oriented) cobordism $f^{-1}(J)$ and the Morse function $f|_{f^{-1}(J)} : f^{-1}(J) \to J$, we see that exactly one end point of J belongs to R. Therefore, $f_*[S(f)]_2$ coincides with the $\mathbf{Z}_2$-homology class represented by ∂R, since f satisfies the normal crossing condition. Thus the result follows. □

Remark 15.5. Proposition 15.4 does not hold for general Thom maps. For example, let $f : S^1 \to S^1$ be a C^∞ homeomorphism such that f is equivalent to the function $x \mapsto x^3$ at a point. Then, f is a Thom map, but $S(f)$ consists exactly of one point.

By Thom [57], the Poincaré dual of $[S(f)]_2 \in H_{p-1}(M; \mathbf{Z}_2)$ coincides with the $(n-p+1)$-st Stiefel-Whitney class $w_{n-p+1}(TM - f^*TN)$ of the difference bundle $TM - f^*TN$. Since every continuous map between smooth manifolds is homotopic to a Boardman map, we obtain the following.

Corollary 15.6. *Let $f : M \to N$ be a continuous map of a smooth closed n-dimensional manifold M into a smooth p-dimensional manifold with $n \geq p$. Suppose either that $n-p$ is even, or that $n - p \equiv 1 \pmod 4$ and M is orientable. Then we have $f_! w_{n-p+1}(TM - f^*TN) = 0 \in H^1(N; \mathbf{Z}_2)$, where*

$$f_! : H^{n-p+1}(M; \mathbf{Z}_2) \to H^1(N; \mathbf{Z}_2)$$

denotes the Gysin homomorphism induced by f.

As another corollary to Proposition 15.4, we have the following.

Corollary 15.7. *Let $f : M \to N$ be a C^∞ stable map of a closed n-dimensional manifold M into a p-dimensional manifold N with $n \geq p$ such that f has only fold points as its singularities. Suppose either that n and p are odd, or that $n - p \equiv 1 \pmod 4$, $p \equiv 1 \pmod 2$ and M is orientable. Then the Euler characteristic of $f(S(f))$ is even.*

The above corollary follows from the fact that in the above situation, $S(f)$ is a $(p-1)$-dimensional closed submanifold of M and that $f|_{S(f)}$ is an immersion with normal crossings (for example, see [16, Chapter III, §4]), together with [37, Corollary 7.3].

Now let $f : M \to N$ be a C^∞ stable map of a closed n-dimensional manifold M into a p-dimensional manifold N such that f has only fold points as its singularities. For $m \geq 0$, we put

$$\Sigma_m(f) = \{y \in N : f^{-1}(y) \cap S(f) \text{ consists exactly of } m \text{ points}\},$$

and for $m \geq 1$, we put

$$\widetilde{\Sigma}_m(f) = f^{-1}(\Sigma_m(f)) \cap S(f).$$

Note that $\Sigma_m(f)$ is a regular submanifold of N of dimension $p - m$, and that $\widetilde{\Sigma}_m(f)$ is a regular submanifold of M of dimension $p - m$.

Then we have the following.

Proposition 15.8. *Let $f : M \to N$ be a C^∞ stable map of a closed n-dimensional manifold M into a p-dimensional manifold N with $n \geq p$ such that f has only fold points as its singularities. Suppose that $n - p$ is even. Then, the $\mathbf{Z}_2$-homology class*

$$[\overline{\Sigma_m(f)}]_2 \in H_{p-m}(\overline{\Sigma_{m-1}(f)}; \mathbf{Z}_2)$$

represented by $\overline{\Sigma_m(f)}$ vanishes for m odd. Furthermore, the $\mathbf{Z}_2$-homology class

$$[\overline{\widetilde{\Sigma}_m(f)}]_2 \in H_{p-m}(\overline{\widetilde{\Sigma}_{m-1}(f)}; \mathbf{Z}_2)$$

represented by $\overline{\widetilde{\Sigma}_m(f)}$ vanishes for m even.

Proof. Take a point $y_0 \in \Sigma_{m-1}(f)$. Let $R \subset \overline{\Sigma_{m-1}(f)}$ be the the closure of the set of points $y \in \Sigma_{m-1}(f)$ such that

$$\frac{\chi(f^{-1}(y)) - \chi(f^{-1}(y_0))}{2}$$

is odd. Then by an argument similar to that in the proof of Proposition 15.4, we see that $[\overline{\Sigma_m(f)}]_2$ coincides with the $\mathbf{Z}_2$-homology class represented by ∂R, since m is odd. Hence the first half of the proposition follows. The second half follows from a similar argument. □

The above proposition shows, for example, that the singular value set $f(S(f))$ of the C^∞ stable map $f : \mathbf{C}P^2 \sharp 2\overline{\mathbf{C}P^2} \to \mathbf{R}^3$ constructed in Chap. 6 cannot be realized as the singular value set of a C^∞ stable map of a closed n-dimensional manifold into $\mathbf{R}^3$ for $n \geq 3$ odd.

16

Further Developments

Since the first version of this book was written as a preprint, there have already been several new developments in the theory of singular fibers of differentiable maps. In this chapter, we briefly present some of them.

16.1 Signature Formula

In Theorem 5.1, we have shown that for a stable map of a closed orientable 4-manifold M into a 3-manifold, the Euler characteristic of M has the same parity as the number of III^8-fibers. In a forthcoming paper [50], which is a joint paper with Takahiro Yamamoto, we will define a sign, $+1$ or -1, for each III^8-fiber of a stable map of an *oriented* 4-manifold into a 3-manifold. This sign is closely related to the orientation of the source 4-manifold M, and is not related to the co-orientation of singular fibers as discussed in Chap. 10. (In fact, the latter is related to the orientation of the target 3-manifold and not the source 4-manifold.) We will show that *the signature of the source oriented* 4*-manifold coincides with the algebraic number of* III^8*-fibers*, where the algebraic number of III^8-fibers means the sum of the signs over all III^8-fibers.

This will be shown as follows. We will first classify the singular fibers of stable maps of orientable 5-manifolds into 4-manifolds by using the method developed in this book. Furthermore, when the source 5-manifold is oriented, we will define an orientation for the 1-dimensional set of points in the target corresponding to III^8-fibers. As a consequence, we will show that if the source 5-manifold is closed and oriented, then the closure of the 1-dimensional set defines an oriented 1-dimensional cycle. This implies that the algebraic number of III^8-fibers for a stable map of an oriented 4-manifold into a 3-manifold is an oriented bordism invariant. In particular, for stable maps into $\mathbf{R}^3$, it gives an oriented cobordism invariant of the source 4-manifold.

On the other hand, we will show that for the example constructed in Chap. 6, the algebraic number of III^8-fibers coincides with the signature of the source 4-manifold $\mathbf{C}P^2 \sharp 2\overline{\mathbf{C}P^2}$. (For this, we will compute the normal Euler number of the surface of definite fold points. See Remark 6.3.) It is well-known that the signature is also an oriented cobordism invariant of the source 4-manifold. More precisely, it is additive under the disjoint union, and two oriented 4-manifolds are oriented cobordant if and only if they have the same signature. Therefore, the result follows when the target 3-manifold is the Euclidean 3-space. For the general 3-manifold case, we use a result of Conner–Floyd [8] about the oriented bordism groups.

Note that then Theorem 5.1 is an easy corollary to this signature theorem.

16.2 Vassiliev Type Invariants for Stable Maps

In [62], Minoru Yamamoto developed a theory of first order Vassiliev type invariants of stable maps by using the topology of singular fibers (for first order Vassiliev type invariants, refer to [59] or [2]). He considered stable maps of orientable 3-manifolds into the plane and classified all the deformations of singular fibers in generic 1- and 2-parameter families of such maps. Using this, he constructed a (portion of) Vassiliev type complex and determined all its cocycles: more precisely, the group of cocycles forms a free abelian group of rank 7. Furthermore, he gave a geometric interpretation to each of the seven generating cocycles. Six of them count the number of singular fibers of codimension 2, namely singular fibers of types $\widetilde{\mathrm{II}}^{00}$, $\widetilde{\mathrm{II}}^{01}$, $\widetilde{\mathrm{II}}^{11}$, $\widetilde{\mathrm{II}}^{3}$, $\widetilde{\mathrm{II}}^{4}$, and $\widetilde{\mathrm{II}}^{a}$ (see Fig. 3.9 for the notation).[1] The other generating cocycle corresponds to the Euler characteristic of the quotient space appearing in the Stein factorization (see Remark 3.12).

This is a first study of such Vassiliev type invariants in the negative codimension case, as long as the author knows.

As an example of another type of a first order Vassiliev type invariant, let us consider the following. Let $f : M \to \mathbf{R}^2$ be a stable map of a closed *oriented* 3-manifold into the plane. Then we can find an oriented 4-manifold W with $\partial W = M$ and a generic map $F : W \to \mathbf{R} \times [0, \infty)$ such that $f = F|_{\partial W} : \partial W \to \mathbf{R} \times \{0\}$. Let us define

$$\sigma(f) = \sigma(W) - ||\mathrm{III}^8(F)|| \in \mathbf{Z},$$

where $\sigma(W)$ is the signature of the oriented 4-manifold W and $||\mathrm{III}^8(F)||$ is the algebraic number of III^8-fibers of the generic map F in the sense of §16.1. Then the signature theorem mentioned in §16.1 implies that $\sigma(f) \in \mathbf{Z}$ does not depend on the choice of W or F so that it is a well-defined isotopy invariant[2]

[1]Note that M. Yamamoto considered stable maps of *orientable* 3-manifolds so that the other codimension 2 singular fibers do not appear.

[2]The author is indebted to Mikio Furuta for the idea of the construction presented here.

of f. This invariant jumps exactly along codimension 1 unstable maps which correspond to a birth/death of a III^8-fiber so that it certainly defines a first order Vassiliev type invariant for such stable maps.

In M. Yamamoto's result [62], this invariant does not appear. This is because he did not use the orientation of the source 3-manifold in order to define the co-orientation of each codimension 1 stratum corresponding to codimension 1 unstable maps in the mapping space.

Any way, the topological structure of singular fibers gives rise to first order Vassiliev type invariants of stable maps in the negative codimension case. There is even a possibility of defining invariants of smooth manifolds using singular fibers of stable maps.

References

1. P.M. Akhmetiev and R.R. Sadykov, *A remark on elimination of singularities for mappings of* 4*-manifolds into* 3*-manifolds*, Topology Appl. **131** (2003), 51–55.
2. V.I. Arnold, *Topological invariants of plane curves and caustics*, University Lecture Series, Vol. 5, Amer. Math. Soc., Providence, RI, 1994.
3. V.I. Arnol'd, V.A. Vasil'ev, V.V. Goryunov and O.V. Lyashko, *Dynamical systems VI. Singularities—Local and global theory*, Encyclopaedia of Mathematical Sciences, 6, Springer-Verlag, Berlin, 1993.
4. M.F. Atiyah, *Bordism and cobordism*, Proc. Cambridge Philos. Soc. **57** (1961), 200–208.
5. J. Boardman, *Singularities of differentiable maps*, Publ. Math. I.H.E.S. **33** (1967), 21–57.
6. Th. Bröcker and K. Jänich, *Introduction to differential topology*, Translated by C.B. and M.J. Thomas, Cambridge Univ. Press, Cambridge, New York, 1982.
7. O. Burlet and G. de Rham, *Sur certaines applications génériques d'une variété close à* 3 *dimensions dans le plan*, Enseign. Math. **20** (1974), 275–292.
8. P.E. Conner and E.E. Floyd, *Differentiable periodic maps*, Ergebnisse der Math. und ihrer Grenzgebiete, Band 33, Springer-Verlag, Berlin, Göttingen, Heidelberg, 1964.
9. J. Damon, *The relation between* C^∞ *and topological stability*, Bol. Soc. Brasil. Mat. **8** (1977), 1–38.
10. J. Damon, *Topological properties of real simple germs, curves, and the nice dimensions* $n > p$, Math. Proc. Camb. Phil. Soc. **89** (1981), 457–472.
11. A. du Plessis, T. Gaffney and L.C. Wilson, *Map-germs determined by their discriminants*, Stratifications, singularities and differential equations, I (Marseille, 1990; Honolulu, HI, 1990), pp. 1–40, Travaux en Cours, 54, Hermann, Paris, 1997.
12. A. du Plessis and T. Wall, *The geometry of topological stability*, London Math. Soc. Monographs, New Series 9, Oxford Science Publ., The Clarendon Press, Oxford Univ. Press, New York, 1995.
13. C. Ehresmann, *Sur l'espaces fibrés différentiables*, C. R. Acad. Sci. Paris **224** (1947), 1611–1612.
14. T. Fukuda, *Topology of folds, cusps and Morin singularities*, in "A Fete of Topology", eds. Y. Matsumoto, T. Mizutani and S. Morita, Academic Press, 1987, pp.331–353.

15. C.G. Gibson, K. Wirthmüller, A.A. du Plessis and E.J.N. Looijenga, *Topological stability of smooth mappings*, Lecture Notes in Math., Vol. 552, Springer-Verlag, Berlin, New York, 1976.
16. M. Golubitsky and V. Guillemin, *Stable mappings and their singularities*, Grad. Texts in Math., Vol. 14, Springer-Verlag, New York, Heidelberg, Berlin, 1973.
17. C.McA. Gordon and J. Luecke, *Knots are determined by their complements*, Bull. Amer. Math. Soc. (N.S.) **20** (1989), 83–87.
18. C.McA. Gordon and J. Luecke, *Knots are determined by their complements*, J. Amer. Math. Soc. **2** (1989), 371–415.
19. A. Haefliger et A. Kosinski, *Un théorème de Thom sur les singularités des applications différentiables*, Séminaire H. Cartan, E. N. S., 1956/57, Exposé no. 8.
20. J.T. Hiratuka, *A fatorização de Stein e o número de singularidades de aplicações estáveis* (in Portuguese), PhD Thesis, University of São Paulo, 2001.
21. K. Ikegami, *Cobordism group of Morse functions on manifolds*, to appear in Hiroshima Math. J.
22. K. Ikegami and O. Saeki, *Cobordism group of Morse functions on surfaces*, J. Math. Soc. Japan **55** (2003), 1081–1094.
23. M.E. Kazaryan, *Hidden singularities and Vassiliev's homology complex of singularity classes* (in Russian), Mat. Sb. **186** (1995), 119–128; English translation in Sb. Math. **186** (1995), 1811–1820.
24. S. Kikuchi and O. Saeki, *Remarks on the topology of folds*, Proc. Amer. Math. Soc. **123** (1995), 905–908.
25. R. Kirby, *The topology of 4-manifolds*, Lecture Notes in Math., Vol. 1374, Springer-Verlag, Berlin, 1989.
26. M. Kobayashi, *Two stable maps of* $\mathbf{C}^2\mathrm{P}$ *into* $\mathbf{R}^3$, Mem. College Ed. Akita Univ. Natur. Sci. **51** (1997), 5–12.
27. M. Kobayashi, *Construction of a stable map of* $\mathbf{C}^2P$ *into 3-space as a stable lift of the moment map*, preprint, 2002.
28. L. Kushner, H. Levine and P. Porto, *Mapping three-manifolds into the plane I*, Bol. Soc. Mat. Mexicana **29** (1984), 11–33.
29. K. Lamotke, *The topology of complex projective varieties after S. Lefschetz*, Topology **20** (1981), 15–51.
30. H. Levine, *Classifying immersions into* $\mathbf{R}^4$ *over stable maps of 3-manifolds into* $\mathbf{R}^2$, Lecture Notes in Math., Vol. 1157, Springer-Verlag, Berlin, 1985.
31. G. Lusztig, J. Milnor and F.P. Peterson, *Semi-characteristics and cobordism*, Topology **8** (1969), 357–359.
32. J.N. Mather, *Stability of* C^∞ *mappings: VI, the nice dimensions*, Lecture Notes in Math., Vol. 192, Springer-Verlag, Berlin, 1971, pp. 207–253.
33. J.W. Milnor, *On manifolds homeomorphic to the 7-sphere*, Ann. of Math. **64** (1956), 399–405.
34. B. Morin, *Formes canoniques des singularités d'une application différentiable*, C. R. Acad. Sci. Paris **260** (1965), 5662–5665, 6503–6506.
35. W. Motta and O. Saeki, *A two colour theorem and the fundamental class of a polyhedron*, in "Real and complex singularities", ed. J.W. Bruce and F. Tari, Proc. of the 5th Workshop on Real and Complex Singularities, Brazil, CHAPMAN & HALL/CRC Research notes in Math. 412, 2000, pp. 94–109.
36. J.J. Nuño Ballesteros and O. Saeki, *On the number of singularities of a generic surface with boundary in a 3-manifold*, Hokkaido Math. J. **27** (1998), 517–544.
37. J.J. Nuño Ballesteros and O. Saeki, *Euler characteristic formulas for simplicial maps*, Math. Proc. Camb. Phil. Soc. **130** (2001), 307–331.

38. T. Ohmoto, *Vassiliev complex for contact classes of real smooth map-germs*, Rep. Fac. Sci. Kagoshima Univ. Math. Phys. Chem. **27** (1994), 1–12.
39. T. Ohmoto, O. Saeki and K. Sakuma, *Self-intersection class for singularities and its application to fold maps*, Trans. Amer. Math. Soc. **355** (2003), 3825–3838.
40. R. Rimányi and A. Szűcs, *Pontrjagin-Thom-type construction for maps with singularities*, Topology **37** (1998), 1177–1191.
41. R. Sadykov, *Elimination of singularities of smooth mappings of* 4*-manifolds into* 3*-manifolds*, preprint, 2003.
42. O. Saeki, *Notes on the topology of folds*, J. Math. Soc. Japan **44** (1992), 551–566.
43. O. Saeki, *Topology of special generic maps of manifolds into Euclidean spaces*, Topology Appl. **49** (1993), 265–293.
44. O. Saeki, *Topology of special generic maps into* $\mathbf{R}^3$, in "Workshop on Real and Complex Singularities", Matemática Contemporânea **5** (1993), 161–186.
45. O. Saeki, *Studying the topology of Morin singularities from a global viewpoint*, Math. Proc. Camb. Phil. Soc. **117** (1995), 223–235.
46. O. Saeki, *Cobordism groups of special generic functions and groups of homotopy spheres*, Japanese J. Math. **28** (2002), 287–297.
47. O. Saeki, *Fold maps on* 4*-manifolds*, Comment. Math. Helv. **78** (2003), 627–647.
48. O. Saeki and K. Sakuma, *On special generic maps into* $\mathbf{R}^3$, Pacific J. Math. **184** (1998), 175–193.
49. O. Saeki and K. Sakuma, *Special generic maps of* 4*-manifolds and compact complex analytic surfaces*, Math. Ann. **313** (1999), 617–633.
50. O. Saeki and T. Yamamoto, *Singular fibers of stable maps and signatures of* 4*-manifolds*, in preparation.
51. K. Sakuma, *On special generic maps of simply connected* $2n$*-manifolds into* $\mathbf{R}^3$, Topology Appl. **50** (1993), 249–261.
52. K. Sakuma, *On the topology of simple fold maps*, Tokyo J. Math. **17** (1994), 21–31.
53. K. Sakuma, *A note on nonremovable cusp singularities*, Hiroshima Math. J. **31** (2001), 461–465.
54. M. Shiota, *Thom's conjecture on triangulations of maps*, Topology **39** (2000), 383–399.
55. A. Szűcs, *Surfaces in* $\mathbf{R}^3$, Bull. London Math. Soc. **18** (1986), 60–66.
56. R. Thom, *Quelques propriétés globales des variétés différentiables*, Comment. Math. Helv. **28** (1954), 17–86.
57. R. Thom, *Les singularités des applications différentiables*, Ann. Inst. Fourier (Grenoble) **6** (1955–56), 43–87.
58. V.A. Vassilyev, *Lagrange and Legendre characteristic classes*, Translated from the Russian, Advanced Studies in Contemporary Mathematics, Vol. 3, Gordon and Breach Science Publishers, New York, 1988.
59. V.A. Vassilyev, *Cohomology of knot spaces*, Theory of singularities and its applications, pp. 23–69, Adv. Soviet Math., Vol. 1, Amer. Math. Soc., Providence, RI, 1990.
60. C.T.C. Wall, *Classification and stability of singularities of smooth maps*, in "Singularity theory" (Trieste, 1991), World Sci. Publishing, River Edge, NJ, 1995, pp. 920–952.
61. K. Wirthmüller, *Singularities determined by their discriminant*, Math. Ann. **252** (1980), 237–245.
62. M. Yamamoto, *First order semi-local invariants of stable maps of 3-manifolds into the plane*, PhD thesis, Kyushu Univ., March 2004.

63. T. Yamamoto, *Classification of singular fibers and its applications* (in Japanese), Master Thesis, Hokkaido Univ., March 2002.
64. T. Yamamoto, *Classification of singular fibres of stable maps from 4-manifolds into 3-manifolds and its applications*, preprint, October 2003.

List of Symbols

$\cong$	a diffeomorphism or an isomorphism between algebraic objects
$\approx$	a homeomorphism
$\sharp$	connected sum
$>>$	the value on the right hand side is sufficiently smaller than the value on the left hand side
$\sim_f$	two points are in the same connected component of a fiber of a map f
$[*]$	(co)homology class represented by a (co)cycle $*$, or τ-cobordism class of $*$
$[X]_2$	fundamental class of a manifold X with respect to the $\mathbf{Z}_2$-coefficient
$\lvert X \rvert$	number of elements in a finite set X
$-X$	manifold X with the reversed orientation for an oriented manifold X
∂X	$\overline{X} \setminus \operatorname{Int} X$ for a subset X of a space, or boundary of a manifold X or a chain X
$\partial\widetilde{\mathfrak{F}}$	set of equivalence classes $\widetilde{\mathfrak{G}}$ of fibers of codimension $\kappa(\widetilde{\mathfrak{F}})+1$ such that $\widetilde{\mathfrak{G}}(f) \subset \overline{\widetilde{\mathfrak{F}}(f)} \setminus \widetilde{\mathfrak{F}}(f)$ for every f
$\mathbf{0}$	equivalence class of a connected regular fiber
$\overline{\mathbf{0}}$	multi-singularity equivalence class of regular fibers
$\mathbf{0}_{\mathrm{e}}(f)$	set of regular values y of a map f with $b_0(f^{-1}(y))$ even
$\mathbf{0}_{\mathrm{o}}(f)$	set of regular values y of a map f with $b_0(f^{-1}(y))$ odd
$\mathbf{0}_{\mathrm{odd}}$	closure of the set of regular values $y \in \mathbf{R}$ of a function f with $b_0(f^{-1}(y))$ odd

I^*	an equivalence class of singular fibers of codimension 1 for stable maps of orientable 4-manifolds into 3-manifolds
$\widetilde{\mathrm{I}}^*$	an equivalence class of singular fibers of codimension 1 for stable maps of 3-manifolds into surfaces
$\bar{\mathrm{I}}^0$	multi-singularity equivalence class of the definite fold mono-germ
$\bar{\mathrm{I}}^1$	multi-singularity equivalence class of the indefinite fold mono-germ
II^*	an equivalence class of singular fibers of codimension 2 for stable maps of orientable 4-manifolds into 3-manifolds
$\widetilde{\mathrm{II}}^*$	an equivalence class of singular fibers of codimension 2 for stable maps of 3-manifolds into surfaces
III^*	an equivalence class of singular fibers of codimension 3 for stable maps of orientable 4-manifolds into 3-manifolds
$\widehat{\mathrm{III}}^8_{\mathrm{e}}$	C^0 equivalence class modulo two circle components of the suspension of $\mathrm{III}^8_{\mathrm{e}}$
$\widehat{\mathrm{III}}^8_{\mathrm{o}}$	C^0 equivalence class modulo two circle components of the suspension of $\mathrm{III}^8_{\mathrm{o}}$
$\alpha(f)$	homology class represented by the cycle $c(f)$ with c being a cocycle representing α
$\alpha(f)^*$	Poincaré dual to $\alpha(f)$
$b_i(X)$	i-th Betti number of a space X
$\mathbf{C}$	complex plane
$c(f)$	closure of the set of points y in the target of a map f such that the fiber over y appears in a cochain c
$\mathcal{C}(f, \varrho)$	$(C^\kappa(f, \varrho), \delta_\kappa(f))_\kappa$
$C^\kappa(f, \varrho)$	cochain group of singular fibers of codimension κ of a map f with respect to an admissible equivalence ϱ
$\mathcal{C}(f^c, \varrho)$	$(C^\kappa(f^c, \varrho), \delta_\kappa\vert_{C^\kappa(f^c,\varrho)})_\kappa$
$C^\kappa(f^c, \varrho)$	subspace of $C^\kappa(\mathcal{T}_{\mathrm{pr}}(n,p), \varrho)$ spanned by the ϱ-equivalence classes of fibers of codimension κ containing no fiber of f
$\mathcal{C}(\Gamma_{n,p}, \varrho^\Gamma)$	universal complex of singular fibers for $\Gamma_{n,p}$-maps with respect to an admissible equivalence ϱ^Γ
$\mathcal{C}(\Gamma' \setminus \Gamma, \varrho^{\Gamma'})$	kernel of $\pi_{\Gamma',\Gamma} : \mathcal{C}(\Gamma', \varrho^{\Gamma'}) \to \mathcal{C}(\Gamma, \varrho^\Gamma)$
$\mathcal{C}(\Gamma, \overline{\varrho}^\Gamma/\varrho^\Gamma)$	cokernel of $\varepsilon_{\varrho^\Gamma,\overline{\varrho}^\Gamma} : \mathcal{C}(\Gamma, \varrho^\Gamma) \to \mathcal{C}(\Gamma, \overline{\varrho}^\Gamma)$

$\mathcal{C}(\Gamma' \setminus \Gamma, \overline{\varrho}^{\Gamma'}/\varrho^{\Gamma'})$ cokernel of $\mathcal{C}(\Gamma' \setminus \Gamma, \varrho^{\Gamma'}) \to \mathcal{C}(\Gamma' \setminus \Gamma, \overline{\varrho}^{\Gamma'})$

$\mathcal{C}(\widetilde{\Gamma}_k, \mathcal{R}_k^{\widetilde{\Gamma}})$ universal complex of singular fibers for $\widetilde{\Gamma}_k$-maps with respect to a stable system of equivalence relations $\mathcal{R}_k^{\widetilde{\Gamma}}$

$C^\kappa(\Gamma^c, \varrho)$ subspace of $C^\kappa(\mathcal{T}_{\mathrm{pr}}(n,p), \varrho)$ spanned by the ϱ-equivalence classes of fibers of codimension κ containing no fiber of a Γ-map

$\mathcal{C}(\mathcal{T}_{\mathrm{pr}}(n,p), \varrho)$ $(C^\kappa(\mathcal{T}_{\mathrm{pr}}(n,p), \varrho), \delta_\kappa)_\kappa$

$C^\kappa(\mathcal{T}_{\mathrm{pr}}(n,p), \varrho)$ cochain group of singular fibers of codimension κ for $\mathcal{T}_{\mathrm{pr}}(n,p)$ with respect to an admissible equivalence ϱ

$\mathcal{C}(\widetilde{\mathcal{T}}_{\mathrm{pr}}(k), \mathcal{R}_k)$ $\varprojlim_p \mathcal{C}(\mathcal{T}_{\mathrm{pr}}(p-k,p), \varrho_{p-k,p})$

$\mathcal{C}(\widetilde{\mathcal{T}}_{\mathrm{pr}}(k), \widehat{\mathcal{R}}_k)$ $(C^\kappa(\widetilde{\mathcal{T}}_{\mathrm{pr}}(k), \widehat{\mathcal{R}}_k), \delta_\kappa)_\kappa$

$C^\kappa(\widetilde{\mathcal{T}}_{\mathrm{pr}}(k), \widehat{\mathcal{R}}_k)$ cochain group of singular fibers of codimension κ for $\widetilde{\mathcal{T}}_{\mathrm{pr}}(k)$ with respect to a stably admissible equivalence $\widehat{\mathcal{R}}_k$

$\chi(X)$ Euler characteristic of a space X

$\chi^*(X)$ semi-characteristic of a manifold X

$\mathcal{CO}(\mathcal{T}_{\mathrm{pr}}(n,p), \varrho^0_{n,p})$ $(CO^\kappa(\mathcal{T}_{\mathrm{pr}}(n,p), \varrho^0_{n,p}), \delta_\kappa)_\kappa$

$CO^\kappa(\mathcal{T}_{\mathrm{pr}}(n,p), \varrho^0_{n,p})$ cochain group of weakly co-orientable singular fibers of codimension κ for $\mathcal{T}_{\mathrm{pr}}(n,p)$ with respect to the C^0 equivalence

$\mathcal{CO}(\widetilde{\mathcal{T}}_{\mathrm{pr}}(k), \mathcal{R}^0_k)$ universal complex of weakly co-orientable singular fibers for $\widetilde{\mathcal{T}}_{\mathrm{pr}}(k)$ with respect to the stable system of C^0 equivalence relations

$\mathcal{CO}(\mathcal{T}_{\mathrm{pr}}(n,p), \varrho_{n,p})$ universal complex of co-orientable singular fibers for $\mathcal{T}_{\mathrm{pr}}(n,p)$ with respect to an admissible equivalence $\varrho_{n,p}$

$\mathcal{CO}(\widetilde{\mathcal{T}}_{\mathrm{pr}}(k), \mathcal{R}_k)$ $\varprojlim_p \mathcal{CO}(\mathcal{T}_{\mathrm{pr}}(p-k,p), \varrho_{p-k,p})$

$\mathrm{Cob}_\tau(N)$ τ-cobordism group of τ-maps of closed manifolds into a manifold $N = N' \times \mathbf{R}$

$\mathrm{Cob}_\tau^{\mathrm{ori}}(N)$ τ-cobordism group of τ-maps of closed oriented manifolds into a manifold $N = N' \times \mathbf{R}$

$\mathbf{C}P^2$ complex projective plane

$\overline{\mathbf{C}P^2}$ complex projective plane with the reversed orientation

$d(V_0, V_1)$ Euler characteristic modulo 2 of an oriented cobordism between oriented $(4k+1)$-dimensional manifolds V_0 and V_1

δ_κ	coboundary map for $\mathcal{C}(\mathcal{T}_{\rm pr}(n,p),\varrho)$, $\mathcal{C}(\widetilde{\mathcal{T}}_{\rm pr}(k),\widehat{\mathcal{R}}_k)$, etc.
$\overline{\delta}_\kappa$	coboundary map for a quotient complex
$\delta_\kappa(f)$	coboundary map for $\mathcal{C}(f,\varrho)$
$\varepsilon_{\varrho,\overline{\varrho}}$	cochain map $\mathcal{C}(\mathcal{T}_{\rm pr}(n,p),\varrho)\to\mathcal{C}(\mathcal{T}_{\rm pr}(n,p),\overline{\varrho})$, etc.
$\varepsilon_{\varrho^\Gamma,\overline{\varrho}^\Gamma}$	cochain map $\mathcal{C}(\Gamma,\varrho^\Gamma)\to\mathcal{C}(\Gamma,\overline{\varrho}^\Gamma)$
$\varepsilon_{\mathcal{R}_k,\overline{\mathcal{R}}_k}$	cochain map $\mathcal{C}(\widetilde{\mathcal{T}}_{\rm pr}(k),\mathcal{R}_k)\to\mathcal{C}(\widetilde{\mathcal{T}}_{\rm pr}(k),\overline{\mathcal{R}}_k)$
$\varepsilon_{\mathcal{R}^{\tilde{\Gamma}'},\overline{\mathcal{R}}^{\tilde{\Gamma}'}}$	cochain map $\mathcal{C}(\widetilde{\Gamma}',\mathcal{R}^{\widetilde{\Gamma}'})\to\mathcal{C}(\widetilde{\Gamma}',\overline{\mathcal{R}}^{\widetilde{\Gamma}'})$
$\overline{f}$	map of the quotient space W_f to the target in the Stein factorization of a map f
$f_!$	Gysin homomorphism induced by a map f
$\mathfrak{F}$, $\mathfrak{G}$, etc.	an equivalence class of (singular) fibers
$\widetilde{\mathfrak{F}}$, $\widetilde{\mathfrak{G}}$, etc.	an equivalence class of fibers with respect to an admissible equivalence
$\widehat{\mathfrak{F}}$, $\widehat{\mathfrak{G}}$, etc.	an equivalence class of fibers with respect to a stably admissible equivalence
$[\widetilde{\mathfrak{F}}:\widetilde{\mathfrak{G}}]$	incidence coefficient
$[\widetilde{\mathfrak{F}}:\widetilde{\mathfrak{G}}]_f$	incidence coefficient with respect to a specific map f
f^c	set of those C^0 equivalence classes of fibers of elements of $\mathcal{T}_{\rm pr}(n,p)$ which contain no fiber of f
$\mathfrak{F}(f)$	set of points y in the target of a map f such that the fiber over y is equivalent to the disjoint union of a fiber of type $\mathfrak{F}$ and some copies of a fiber of the trivial circle bundle
$\lVert\widetilde{\mathfrak{F}}(f)\rVert$	algebraic number of fibers of type $\widetilde{\mathfrak{F}}$ for a map f
$\mathfrak{F}_{\rm e}$	equivalence class modulo two circle components represented by elements of $\mathfrak{F}_n$ with n even
$\mathfrak{F}_{\rm e}(f)$	subset of $\mathfrak{F}(f)$ consisting of the points y with $b_0(f^{-1}(y))$ even
$F_{(\ell)}$	surface obtained from a closed surface F by taking off ℓ open disks whose closures do not intersect each other
$\mathfrak{F}_n$	equivalence class of a fiber consisting of a fiber of type $\mathfrak{F}$ and some copies of a fiber of the trivial circle bundle such that the total number of connected components is equal to n
$\mathfrak{F}_{\rm o}$	equivalence class modulo two circle components represented by elements of $\mathfrak{F}_n$ with n odd

$\mathfrak{F}_o(f)$	subset of $\mathfrak{F}(f)$ consisting of the points y with $b_0(f^{-1}(y))$ odd
$\Gamma = \Gamma_{n,p}$	an ascending set of C^0 equivalence classes of fibers of elements of $\mathcal{T}_{\mathrm{pr}}(n,p)$, or set of all Γ-maps
$\widetilde{\Gamma} = \widetilde{\Gamma}_k = \bigcup_{p-n=k} \Gamma_{n,p}$	a set of C^0 equivalence classes of fibers of proper Thom maps of codimension k such that each $\Gamma_{n,p}$ is an ascending set of C^0 equivalence classes of fibers of elements of $\mathcal{T}_{\mathrm{pr}}(n,p)$, and that $\widetilde{\Gamma}$ is closed under suspension, or set of all $\widetilde{\Gamma}_k$-maps
$\Gamma^* = \Gamma^*_{n,p}$	set of all C^0 equivalence classes of fibers of elements of $\Gamma_{n,p}$, or set of all $\Gamma^*_{n,p}$-maps
$\widetilde{\Gamma}^* = \widetilde{\Gamma}^*_k$	set of all C^0 equivalence classes of fibers of elements of $\widetilde{\Gamma}_k$ and their suspensions
$H^\kappa(*)$	κ-th cohomology group of a relevant complex
id_X	identity map of a space X
$\mathrm{Ind}_f(y)$	index of the image of a swallowtail (or a cross cap) y of a map f
$\mathrm{Int}\, V$	interior of a subspace V of a space, or interior of a manifold V
K^2	Klein bottle
$\kappa(\mathfrak{F})$	codimension of an equivalence class $\mathfrak{F}$ of fibers
$\kappa(\widetilde{\mathfrak{F}})$	codimension of an equivalence class $\widetilde{\mathfrak{F}}$ of fibers with respect to an admissible equivalence
$\kappa(\widehat{\mathfrak{F}})$	codimension of an equivalence class $\widehat{\mathfrak{F}}$ of fibers with respect to a stably admissible equivalence
$\mathcal{M}$	a Whitney stratification of a manifold M
$\mathcal{M}_{\mathrm{pr}}(n,p)$	set of all proper Morin maps in $\mathcal{T}_{\mathrm{pr}}(n,p)$ which satisfy the normal crossing condition
$\mathcal{M}_{\mathrm{pr}}(n,p)^{\mathrm{ori}}$	subset of $\mathcal{M}_{\mathrm{pr}}(n,p)$ consisting of those maps whose source manifolds are orientable
$\widetilde{\mathcal{M}}_{\mathrm{pr}}(k)$	$\bigcup_{p-n=k} \mathcal{M}_{\mathrm{pr}}(n,p)$
$\widetilde{\mathcal{M}}_{\mathrm{pr}}(k)^{\mathrm{ori}}$	$\bigcup_{p-n=k} \mathcal{M}_{\mathrm{pr}}(n,p)^{\mathrm{ori}}$
$\mathcal{N}$	a Whitney stratification of a manifold N
$n_{\widetilde{\mathfrak{F}}}(\widetilde{\mathfrak{G}})$	number of $\widetilde{\mathfrak{G}}$-branches in a neighborhood of a top dimensional $\widetilde{\mathfrak{F}}$-stratum

φ_f	homomorphism $H^\kappa(\Gamma_{n,p}, \varrho^\Gamma_{n,p}) \to H^\kappa(N; \mathbf{Z}_2)$ induced by a Thom map $f : M \to N$
$\widetilde{\varphi}_f$	$\varphi_f \circ \Phi^\kappa_{n,p*} : H^\kappa(\widetilde{\Gamma}_k, \mathcal{R}^{\widetilde{\Gamma}}_k) \to H^\kappa(N; \mathbf{Z}_2)$
Φ	cochain map $\mathcal{C}(\widetilde{\mathcal{T}}_{\mathrm{pr}}(k), \widehat{\mathcal{R}}_k) \to \varprojlim_p \mathcal{C}(\mathcal{T}_{\mathrm{pr}}(p-k,p), \varrho_{p-k,p}) = \mathcal{C}(\widetilde{\mathcal{T}}_{\mathrm{pr}}(k), \mathcal{R}_k)$
Φ_κ	homomorphism $\mathrm{Cob}_\tau(N) \to \mathrm{Hom}\,(\mathrm{Im}\, s_{\kappa*}, H^\kappa(N; \mathbf{Z}_2))$ defined by $\Phi_\kappa([f]) = \varphi_f\|_{\mathrm{Im}\, s_{\kappa*}}$ for a τ-map f
$\Phi_{n,p}$	$\{\Phi^\kappa_{n,p}\}_\kappa$
$\Phi^\kappa_{n,p}$	homomorphism induced by the projection $C^\kappa(\widetilde{\mathcal{T}}_{\mathrm{pr}}(k), \mathcal{R}_k) \to C^\kappa(\mathcal{T}_{\mathrm{pr}}(n,p), \varrho_{n,p})$, etc.
$\pi_{\Gamma',\Gamma}$	cochain map $\mathcal{C}(\Gamma', \varrho^{\Gamma'}) \to \mathcal{C}(\Gamma, \varrho^{\Gamma})$ induced by the projection
$\pi_{\widetilde{\Gamma}',\widetilde{\Gamma}}$	cochain map $\mathcal{C}(\widetilde{\Gamma}', \mathcal{R}^{\widetilde{\Gamma}'}_k) \to \mathcal{C}(\widetilde{\Gamma}, \mathcal{R}^{\widetilde{\Gamma}}_k)$ induced by the projection
q_f	quotient map in the Stein factorization of a map f
$\mathbf{R}$	real line
$\mathcal{R}_k$ or $\overline{\mathcal{R}}_k$	a stable system $\{\varrho_{p-k,p}\}_p$ (resp. $\{\overline{\varrho}_{p-k,p}\}_p$) of admissible equivalence relations for the fibers of codimension k proper Thom maps
$\mathcal{R}_k \le \overline{\mathcal{R}}_k$	$\mathcal{R}_k$ is weaker than $\overline{\mathcal{R}}_k$
$\widehat{\mathcal{R}}_k$	an equivalence relation among the fibers of proper Thom maps of codimension k, or of maps in $\widetilde{\Gamma}$
$\mathcal{R}^0_k$	$\{\varrho^0_{p-k,p}\}_p$
$\widehat{\mathcal{R}}^0_k$	stable C^0 equivalence relation among the fibers of proper Thom maps of codimension k
$\mathcal{R}^0_{-1}(m)$	$\{\varrho^0_{p+1,p}(m)\}_{p\ge 0}$
$\varrho = \varrho_{n,p}$ or $\overline{\varrho} = \overline{\varrho}_{n,p}$	an admissible equivalence relation among the fibers of proper Thom maps between manifolds of dimensions n and p
$\varrho_{n,p} \le \overline{\varrho}_{n,p}$	$\varrho_{n,p}$ is weaker than $\overline{\varrho}_{n,p}$
$\varrho^0_{n,p}$	C^0 equivalence relation among the fibers of proper Thom maps between manifolds of dimensions n and p
$\varrho^0_{p+1,p}(m)$	C^0 equivalence modulo m circle components

$\varrho^{\Gamma} = \varrho^{\Gamma}_{n,p}$, $\varrho^{\Gamma'}, \overline{\varrho}^{\Gamma}, \overline{\varrho}^{\Gamma'}$, etc.	an admissible equivalence relation among the fibers of Γ-maps, etc.
$\varrho^{\mathrm{ms}}_{n,p}$	multi-singularity equivalence
$\mathcal{R}_k$	a system of equivalence relations $\{\varrho_{p-k,p}\}_p$
$\mathcal{R}^0_k$	system of the C^0 equivalence relations $\{\varrho^0_{p-k,p}\}_p$
$\mathcal{R}^{\widetilde{\Gamma}}_k$	a system of equivalence relations $\{\varrho^{\Gamma_{p-k,p}}_{p-k,p}\}_p$ such that each $\varrho^{\Gamma_{p-k,p}}_{p-k,p}$ is an admissible equivalence relation among the fibers of $\Gamma_{p-k,p}$-maps
$\mathcal{R}^{\tau}_k$	a stable system $\{\varrho^{\tau}_{p-k,p}\}_p$ of admissible equivalence relations for the fibers of elements of $\widetilde{\tau}(k)$
$\mathbf{R}P^2$	real projective plane
$S(f)$	singular set of a smooth map f
s_{κ}	homomorphism induced by the suspension $C^{\kappa}(\mathcal{T}_{\mathrm{pr}}(n+\ell, p+\ell), \varrho_{n+\ell,p+\ell}) \to C^{\kappa}(\mathcal{T}_{\mathrm{pr}}(n,p), \varrho_{n,p})$
$\mathcal{S}^0_{\mathrm{pr}}(n,p)$	set of all C^0 stable maps which are elements of $\mathcal{T}_{\mathrm{pr}}(n,p)$
$\mathcal{S}^0_{\mathrm{pr}}(n,p)^{\mathrm{ori}}$	subset of $S^0_{\mathrm{pr}}(n,p)$ consisting of those maps whose source manifolds are orientable
$\widetilde{\mathcal{S}}^0_{\mathrm{pr}}(k)$	$\bigcup_{p-n=k} \mathcal{S}^0_{\mathrm{pr}}(n,p)$
$\mathcal{S}^{\infty}_{\mathrm{pr}}(n,p)$	set of all proper C^{∞} stable maps between manifolds of dimensions n and p
$\widetilde{\mathcal{S}}^{\infty}_{\mathrm{pr}}(k)$	$\bigcup_{p-n=k} \mathcal{S}^{\infty}_{\mathrm{pr}}(n,p)$
$\Sigma\eta$	suspension of a map-germ η
Σf	suspension of a map f
$\Sigma_m(f)$	set of points y in the target of a map f such that $f^{-1}(y) \cap S(f)$ consists exactly of m points
$\widetilde{\Sigma}_m(f)$	$f^{-1}(\Sigma_m(f)) \cap S(f)$ for a map f
T^2	2-dimensional torus
$\mathcal{T}_{\mathrm{pr}}(n,p)$	set of all proper Thom maps between manifolds of dimensions n and p
$\widetilde{\mathcal{T}}_{\mathrm{pr}}(k)$	$\bigcup_{p-n=k} \mathcal{T}_{\mathrm{pr}}(n,p)$
τ	an ascending set of singularity types

$\tau(n,p)$	set of all proper Thom maps between manifolds of dimensions n and p which are τ-maps
$\widetilde{\tau}(k)$	$\bigcup_p \tau(p-k,p)$
$\tau^0(n,p)$	set of C^0 stable maps in $\mathcal{T}_{\mathrm{pr}}(n,p)$ which are τ-maps
$\tau^0(n,p)^{\mathrm{ori}}$	set of all C^0 equivalence classes of fibers for proper C^0 stable τ-maps in $\mathcal{T}_{\mathrm{pr}}(n,p)$ of orientable n-dimensional manifolds
W_f	quotient space in the Stein factorization of a map f
w_i	i-th Stiefel-Whitney class
$\mathbf{Z}$	set of all integers
$\mathbf{Z}_2$	prime field of characteristic 2

Index

Printing and Binding: Strauss GmbH, Mörlenbach

Lecture Notes in Mathematics

For information about Vols. 1–1679
please contact your bookseller or Springer

Vol. 1680: K.-Z. Li, F. Oort, Moduli of Supersingular Abelian Varieties. V, 116 pages. 1998.

Vol. 1681: G. J. Wirsching, The Dynamical System Generated by the 3n+1 Function. VII, 158 pages. 1998.

Vol. 1682: H.-D. Alber, Materials with Memory. X, 166 pages. 1998.

Vol. 1683: A. Pomp, The Boundary-Domain Integral Method for Elliptic Systems. XVI, 163 pages. 1998.

Vol. 1684: C. A. Berenstein, P. F. Ebenfelt, S. G. Gindikin, S. Helgason, A. E. Tumanov, Integral Geometry, Radon Transforms and Complex Analysis. Firenze, 1996. Editors: E. Casadio Tarabusi, M. A. Picardello, G. Zampieri. VII, 160 pages. 1998.

Vol. 1685: S. König, A. Zimmermann, Derived Equivalences for Group Rings. X, 146 pages. 1998.

Vol. 1686: J. Azéma, M. Émery, M. Ledoux, M. Yor (Eds.), Séminaire de Probabilités XXXII. VI, 440 pages. 1998.

Vol. 1687: F. Bornemann, Homogenization in Time of Singularly Perturbed Mechanical Systems. XII, 156 pages. 1998.

Vol. 1688: S. Assing, W. Schmidt, Continuous Strong Markov Processes in Dimension One. XII, 137 page. 1998.

Vol. 1689: W. Fulton, P. Pragacz, Schubert Varieties and Degeneracy Loci. XI, 148 pages. 1998.

Vol. 1690: M. T. Barlow, D. Nualart, Lectures on Probability Theory and Statistics. Editor: P. Bernard. VIII, 237 pages. 1998.

Vol. 1691: R. Bezrukavnikov, M. Finkelberg, V. Schechtman, Factorizable Sheaves and Quantum Groups. X, 282 pages. 1998.

Vol. 1692: T. M. W. Eyre, Quantum Stochastic Calculus and Representations of Lie Superalgebras. IX, 138 pages. 1998.

Vol. 1694: A. Braides, Approximation of Free-Discontinuity Problems. XI, 149 pages. 1998.

Vol. 1695: D. J. Hartfiel, Markov Set-Chains. VIII, 131 pages. 1998.

Vol. 1696: E. Bouscaren (Ed.): Model Theory and Algebraic Geometry. XV, 211 pages. 1998.

Vol. 1697: B. Cockburn, C. Johnson, C.-W. Shu, E. Tadmor, Advanced Numerical Approximation of Nonlinear Hyperbolic Equations. Cetraro, Italy, 1997. Editor: A. Quarteroni. VII, 390 pages. 1998.

Vol. 1698: M. Bhattacharjee, D. Macpherson, R. G. Möller, P. Neumann, Notes on Infinite Permutation Groups. XI, 202 pages. 1998.

Vol. 1699: A. Inoue,Tomita-Takesaki Theory in Algebras of Unbounded Operators. VIII, 241 pages. 1998.

Vol. 1700: W. A. Woyczyński, Burgers-KPZ Turbulence, XI, 318 pages. 1998.

Vol. 1701: Ti-Jun Xiao, J. Liang, The Cauchy Problem of Higher Order Abstract Differential Equations, XII, 302 pages. 1998.

Vol. 1702: J. Ma, J. Yong, Forward-Backward Stochastic Differential Equations and Their Applications. XIII, 270 pages. 1999.

Vol. 1703: R. M. Dudley, R. Norvaiša, Differentiability of Six Operators on Nonsmooth Functions and p-Variation. VIII, 272 pages. 1999.

Vol. 1704: H. Tamanoi, Elliptic Genera and Vertex Operator Super-Algebras. VI, 390 pages. 1999.

Vol. 1705: I. Nikolaev, E. Zhuzhoma, Flows in 2-dimensional Manifolds. XIX, 294 pages. 1999.

Vol. 1706: S. Yu. Pilyugin, Shadowing in Dynamical Systems. XVII, 271 pages. 1999.

Vol. 1707: R. Pytlak, Numerical Methods for Optimal Control Problems with State Constraints. XV, 215 pages. 1999.

Vol. 1708: K. Zuo, Representations of Fundamental Groups of Algebraic Varieties. VII, 139 pages. 1999.

Vol. 1709: J. Azéma, M. Émery, M. Ledoux, M. Yor (Eds.), Séminaire de Probabilités XXXIII. VIII, 418 pages. 1999.

Vol. 1710: M. Koecher, The Minnesota Notes on Jordan Algebras and Their Applications. IX, 173 pages. 1999.

Vol. 1711: W. Ricker, Operator Algebras Generated by Commuting Projections: A Vector Measure Approach. XVII, 159 pages. 1999.

Vol. 1712: N. Schwartz, J. J. Madden, Semi-algebraic Function Rings and Reflectors of Partially Ordered Rings. XI, 279 pages. 1999.

Vol. 1713: F. Bethuel, G. Huisken, S. Müller, K. Steffen, Calculus of Variations and Geometric Evolution Problems. Cetraro, 1996. Editors: S. Hildebrandt, M. Struwe. VII, 293 pages. 1999.

Vol. 1714: O. Diekmann, R. Durrett, K. P. Hadeler, P. K. Maini, H. L. Smith, Mathematics Inspired by Biology. Martina Franca, 1997. Editors: V. Capasso, O. Diekmann. VII, 268 pages. 1999.

Vol. 1715: N. V. Krylov, M. Röckner, J. Zabczyk, Stochastic PDE's and Kolmogorov Equations in Infinite Dimensions. Cetraro, 1998. Editor: G. Da Prato. VIII, 239 pages. 1999.

Vol. 1716: J. Coates, R. Greenberg, K. A. Ribet, K. Rubin, Arithmetic Theory of Elliptic Curves. Cetraro, 1997. Editor: C. Viola. VIII, 260 pages. 1999.

Vol. 1717: J. Bertoin, F. Martinelli, Y. Peres, Lectures on Probability Theory and Statistics. Saint-Flour, 1997. Editor: P. Bernard. IX, 291 pages. 1999.

Vol. 1718: A. Eberle, Uniqueness and Non-Uniqueness of Semigroups Generated by Singular Diffusion Operators. VIII, 262 pages. 1999.

Vol. 1719: K. R. Meyer, Periodic Solutions of the N-Body Problem. IX, 144 pages. 1999.

Vol. 1720: D. Elworthy, Y. Le Jan, X-M. Li, On the Geometry of Diffusion Operators and Stochastic Flows. IV, 118 pages. 1999.

Vol. 1721: A. Iarrobino, V. Kanev, Power Sums, Gorenstein Algebras, and Determinantal Loci. XXVII, 345 pages. 1999.

Vol. 1722: R. McCutcheon, Elemental Methods in Ergodic Ramsey Theory. VI, 160 pages. 1999.

Vol. 1723: J. P. Croisille, C. Lebeau, Diffraction by an Immersed Elastic Wedge. VI, 134 pages. 1999.

Vol. 1724: V. N. Kolokoltsov, Semiclassical Analysis for Diffusions and Stochastic Processes. VIII, 347 pages. 2000.

Vol. 1725: D. A. Wolf-Gladrow, Lattice-Gas Cellular Automata and Lattice Boltzmann Models. IX, 308 pages. 2000.

Vol. 1726: V. Marić, Regular Variation and Differential Equations. X, 127 pages. 2000.

Vol. 1727: P. Kravanja M. Van Barel, Computing the Zeros of Analytic Functions. VII, 111 pages. 2000.

Vol. 1728: K. Gatermann Computer Algebra Methods for Equivariant Dynamical Systems. XV, 153 pages. 2000.

Vol. 1729: J. Azéma, M. Émery, M. Ledoux, M. Yor (Eds.) Séminaire de Probabilités XXXIV. VI, 431 pages. 2000.

Vol. 1730: S. Graf, H. Luschgy, Foundations of Quantization for Probability Distributions. X, 230 pages. 2000.

Vol. 1731: T. Hsu, Quilts: Central Extensions, Braid Actions, and Finite Groups. XII, 185 pages. 2000.

Vol. 1732: K. Keller, Invariant Factors, Julia Equivalences and the (Abstract) Mandelbrot Set. X, 206 pages. 2000.

Vol. 1733: K. Ritter, Average-Case Analysis of Numerical Problems. IX, 254 pages. 2000.

Vol. 1734: M. Espedal, A. Fasano, A. Mikelić, Filtration in Porous Media and Industrial Applications. Cetraro 1998. Editor: A. Fasano. 2000.

Vol. 1735: D. Yafaev, Scattering Theory: Some Old and New Problems. XVI, 169 pages. 2000.

Vol. 1736: B. O. Turesson, Nonlinear Potential Theory and Weighted Sobolev Spaces. XIV, 173 pages. 2000.

Vol. 1737: S. Wakabayashi, Classical Microlocal Analysis in the Space of Hyperfunctions. VIII, 367 pages. 2000.

Vol. 1738: M. Émery, A. Nemirovski, D. Voiculescu, Lectures on Probability Theory and Statistics. XI, 356 pages. 2000.

Vol. 1739: R. Burkard, P. Deuflhard, A. Jameson, J.-L. Lions, G. Strang, Computational Mathematics Driven by Industrial Problems. Martina Franca, 1999. Editors: V. Capasso, H. Engl, J. Periaux. VII, 418 pages. 2000.

Vol. 1740: B. Kawohl, O. Pironneau, L. Tartar, J.-P. Zolesio, Optimal Shape Design. Tróia, Portugal 1999. Editors: A. Cellina, A. Ornelas. IX, 388 pages. 2000.

Vol. 1741: E. Lombardi, Oscillatory Integrals and Phenomena Beyond all Algebraic Orders. XV, 413 pages. 2000.

Vol. 1742: A. Unterberger, Quantization and Nonholomorphic Modular Forms.VIII, 253 pages. 2000.

Vol. 1743: L. Habermann, Riemannian Metrics of Constant Mass and Moduli Spaces of Conformal Structures. XII, 116 pages. 2000.

Vol. 1744: M. Kunze, Non-Smooth Dynamical Systems. X, 228 pages. 2000.

Vol. 1745: V. D. Milman, G. Schechtman (Eds.), Geometric Aspects of Functional Analysis. Israel Seminar 1999-2000. VIII, 289 pages. 2000.

Vol. 1746: A. Degtyarev, I. Itenberg, V. Kharlamov, Real Enriques Surfaces. XVI, 259 pages. 2000.

Vol. 1747: L. W. Christensen, Gorenstein Dimensions. VIII, 204 pages. 2000.

Vol. 1748: M. Ruzicka, Electrorheological Fluids: Modeling and Mathematical Theory. XV, 176 pages. 2001.

Vol. 1749: M. Fuchs, G. Seregin, Variational Methods for Problems from Plasticity Theory and for Generalized Newtonian Fluids. VI, 269 pages. 2001.

Vol. 1750: B. Conrad, Grothendieck Duality and Base Change. X, 296 pages. 2001.

Vol. 1751: N. J. Cutland, Loeb Measures in Practice: Recent Advances. XI, 111 pages. 2001.

Vol. 1752: Y. V. Nesterenko, P. Philippon, Introduction to Algebraic Independence Theory. XIII, 256 pages. 2001.

Vol. 1753: A. I. Bobenko, U. Eitner, Painlevé Equations in the Differential Geometry of Surfaces. VI, 120 pages. 2001.

Vol. 1754: W. Bertram, The Geometry of Jordan and Lie Structures. XVI, 269 pages. 2001.

Vol. 1755: J. Azéma, M. Émery, M. Ledoux, M. Yor (Eds.), Séminaire de Probabilités XXXV. VI, 427 pages. 2001.

Vol. 1756: P. E. Zhidkov, Korteweg de Vries and Nonlinear Schrödinger Equations: Qualitative Theory. VII, 147 pages. 2001.

Vol. 1757: R. R. Phelps, Lectures on Choquet's Theorem. VII, 124 pages. 2001.

Vol. 1758: N. Monod, Continuous Bounded Cohomology of Locally Compact Groups. X, 214 pages. 2001.

Vol. 1759: Y. Abe, K. Kopfermann, Toroidal Groups. VIII, 133 pages. 2001.

Vol. 1760: D. Filipović, Consistency Problems for Heath-Jarrow-Morton Interest Rate Models. VIII, 134 pages. 2001.

Vol. 1761: C. Adelmann, The Decomposition of Primes in Torsion Point Fields. VI, 142 pages. 2001.

Vol. 1762: S. Cerrai, Second Order PDE's in Finite and Infinite Dimension. IX, 330 pages. 2001.

Vol. 1763: J.-L. Loday, A. Frabetti, F. Chapoton, F. Goichot, Dialgebras and Related Operads. IV, 132 pages. 2001.

Vol. 1764: A. Cannas da Silva, Lectures on Symplectic Geometry. XII, 217 pages. 2001.

Vol. 1765: T. Kerler, V. V. Lyubashenko, Non-Semisimple Topological Quantum Field Theories for 3-Manifolds with Corners. VI, 379 pages. 2001.

Vol. 1766: H. Hennion, L. Hervé, Limit Theorems for Markov Chains and Stochastic Properties of Dynamical Systems by Quasi-Compactness. VIII, 145 pages. 2001.

Vol. 1767: J. Xiao, Holomorphic Q Classes. VIII, 112 pages. 2001.

Vol. 1768: M.J. Pflaum, Analytic and Geometric Study of Stratified Spaces. VIII, 230 pages. 2001.

Vol. 1769: M. Alberich-Carramiñana, Geometry of the Plane Cremona Maps. XVI, 257 pages. 2002.

Vol. 1770: H. Gluesing-Luerssen, Linear Delay-Differential Systems with Commensurate Delays: An Algebraic Approach. VIII, 176 pages. 2002.

Vol. 1771: M. Émery, M. Yor (Eds.), Séminaire de Probabilités 1967-1980. A Selection in Martingale Theory. IX, 553 pages. 2002.

Vol. 1772: F. Burstall, D. Ferus, K. Leschke, F. Pedit, U. Pinkall, Conformal Geometry of Surfaces in S^4. VII, 89 pages. 2002.

Vol. 1773: Z. Arad, M. Muzychuk, Standard Integral Table Algebras Generated by a Non-real Element of Small Degree. X, 126 pages. 2002.

Vol. 1774: V. Runde, Lectures on Amenability. XIV, 296 pages. 2002.

Vol. 1775: W. H. Meeks, A. Ros, H. Rosenberg, The Global Theory of Minimal Surfaces in Flat Spaces. Martina Franca 1999. Editor: G. P. Pirola. X, 117 pages. 2002.

Vol. 1776: K. Behrend, C. Gomez, V. Tarasov, G. Tian, Quantum Comohology. Cetraro 1997. Editors: P. de Bartolomeis, B. Dubrovin, C. Reina. VIII, 319 pages. 2002.

Vol. 1777: E. García-Río, D. N. Kupeli, R. Vázquez-Lorenzo, Osserman Manifolds in Semi-Riemannian Geometry. XII, 166 pages. 2002.

Vol. 1778: H. Kiechle, Theory of K-Loops. X, 186 pages. 2002.

Vol. 1779: I. Chueshov, Monotone Random Systems. VIII, 234 pages. 2002.

Vol. 1780: J. H. Bruinier, Borcherds Products on O(2,1) and Chern Classes of Heegner Divisors. VIII, 152 pages. 2002.

Vol. 1781: E. Bolthausen, E. Perkins, A. van der Vaart, Lectures on Probability Theory and Statistics. Ecole d' Eté de Probabilités de Saint-Flour XXIX-1999. Editor: P. Bernard. VIII, 466 pages. 2002.

Vol. 1782: C.-H. Chu, A. T.-M. Lau, Harmonic Functions on Groups and Fourier Algebras. VII, 100 pages. 2002.

Vol. 1783: L. Grüne, Asymptotic Behavior of Dynamical and Control Systems under Perturbation and Discretization. IX, 231 pages. 2002.

Vol. 1784: L.H. Eliasson, S. B. Kuksin, S. Marmi, J.-C. Yoccoz, Dynamical Systems and Small Divisors. Cetraro, Italy 1998. Editors: S. Marmi, J.-C. Yoccoz. VIII, 199 pages. 2002.

Vol. 1785: J. Arias de Reyna, Pointwise Convergence of Fourier Series. XVIII, 175 pages. 2002.

Vol. 1786: S. D. Cutkosky, Monomialization of Morphisms from 3-Folds to Surfaces. V, 235 pages. 2002.

Vol. 1787: S. Caenepeel, G. Militaru, S. Zhu, Frobenius and Separable Functors for Generalized Module Categories and Nonlinear Equations. XIV, 354 pages. 2002.

Vol. 1788: A. Vasil'ev, Moduli of Families of Curves for Conformal and Quasiconformal Mappings.IX, 211 pages. 2002.

Vol. 1789: Y. Sommerhäuser, Yetter-Drinfel'd Hopf algebras over groups of prime order. V, 157 pages. 2002.

Vol. 1790: X. Zhan, Matrix Inequalities. VII, 116 pages. 2002.

Vol. 1791: M. Knebusch, D. Zhang, Manis Valuations and Prüfer Extensions I: A new Chapter in Commutative Algebra. VI, 267 pages. 2002.

Vol. 1792: D. D. Ang, R. Gorenflo, V. K. Le, D. D. Trong, Moment Theory and Some Inverse Problems in Potential Theory and Heat Conduction. VIII, 183 pages. 2002.

Vol. 1793: J. Cortés Monforte, Geometric, Control and Numerical Aspects of Nonholonomic Systems. XV, 219 pages. 2002.

Vol. 1794: N. Pytheas Fogg, Substitution in Dynamics, Arithmetics and Combinatorics. Editors: V. Berthé, S. Ferenczi, C. Mauduit, A. Siegel. XVII, 402 pages. 2002.

Vol. 1795: H. Li, Filtered-Graded Transfer in Using Noncommutative Gröbner Bases. IX, 197 pages. 2002.

Vol. 1796: J.M. Melenk, hp-Finite Element Methods for Singular Perturbations. XIV, 318 pages. 2002.

Vol. 1797: B. Schmidt, Characters and Cyclotomic Fields in Finite Geometry. VIII, 100 pages. 2002.

Vol. 1798: W.M. Oliva, Geometric Mechanics. XI, 270 pages. 2002.

Vol. 1799: H. Pajot, Analytic Capacity, Rectifiability, Menger Curvature and the Cauchy Integral. XII,119 pages. 2002.

Vol. 1800: O. Gabber, L. Ramero, Almost Ring Theory. VI, 307 pages. 2003.

Vol. 1801: J. Azéma, M. Émery, M. Ledoux, M. Yor (Eds.), Séminaire de Probabilités XXXVI. VIII, 499 pages. 2003.

Vol. 1802: V. Capasso, E. Merzbach, B.G. Ivanoff, M. Dozzi, R. Dalang, T. Mountford, Topics in Spatial Stochastic Processes. Martina Franca, Italy 2001. Editor: E. Merzbach. VIII, 253 pages. 2003.

Vol. 1803: G. Dolzmann, Variational Methods for Crystalline Microstructure - Analysis and Computation. VIII, 212 pages. 2003.

Vol. 1804: I. Cherednik, Ya. Markov, R. Howe, G. Lusztig, Iwahori-Hecke Algebras and their Representation Theory. Martina Franca, Italy 1999. Editors: V. Baldoni, D. Barbasch. X, 103 pages. 2003.

Vol. 1805: F. Cao, Geometric Curve Evolution and Image Processing. X, 187 pages. 2003.

Vol. 1806: H. Broer, I. Hoveijn. G. Lunther, G. Vegter, Bifurcations in Hamiltonian Systems. Computing Singularities by Gröbner Bases. XIV, 169 pages. 2003.

Vol. 1807: V. D. Milman, G. Schechtman (Eds.), Geometric Aspects of Functional Analysis. Israel Seminar 2000-2002. VIII, 429 pages. 2003.

Vol. 1808: W. Schindler, Measures with Symmetry Properties.IX, 167 pages. 2003.

Vol. 1809: O. Steinbach, Stability Estimates for Hybrid Coupled Domain Decomposition Methods. VI, 120 pages. 2003.

Vol. 1810: J. Wengenroth, Derived Functors in Functional Analysis. VIII, 134 pages. 2003.

Vol. 1811: J. Stevens, Deformations of Singularities. VII, 157 pages. 2003.

Vol. 1812: L. Ambrosio, K. Deckelnick, G. Dziuk, M. Mimura, V. A. Solonnikov, H. M. Soner, Mathematical Aspects of Evolving Interfaces. Madeira, Funchal, Portugal 2000. Editors: P. Colli, J. F. Rodrigues. X, 237 pages. 2003.

Vol. 1813: L. Ambrosio, L. A. Caffarelli, Y. Brenier, G. Buttazzo, C. Villani, Optimal Transportation and its Applications. Martina Franca, Italy 2001. Editors: L. A. Caffarelli, S. Salsa. X, 164 pages. 2003.

Vol. 1814: P. Bank, F. Baudoin, H. Föllmer, L.C.G. Rogers, M. Soner, N. Touzi, Paris-Princeton Lectures on Mathematical Finance. X,172 pages. 2003.

Vol. 1815: A. M. Vershik (Ed.), Asymptotic Combinatorics with Applications to Mathematical Physics. St. Petersburg, Russia 2001. IX, 246 pages. 2003.

Vol. 1816: S. Albeverio, W. Schachermayer, M. Talagrand, Lectures on Probability Theory and Statistics. Ecole d'Eté de Probabilités de Saint-Flour XXX-2000. Editor: P. Bernard. VIII, 296 pages. 2003.

Vol. 1817: E. Koelink (Ed.), Orthogonal Polynomials and Special Functions. Leuven 2002. X, 249 pages. 2003.

Vol. 1818: M. Bildhauer, Convex Variational Problems with Linear, nearly Linear and/or Anisotropic Growth Conditions. X, 217 pages. 2003.

Vol. 1819: D. Masser, Yu. V. Nesterenko, H. P. Schlickewei, W. M. Schmidt, M. Waldschmidt, Diophantine Approximation. Cetraro, Italy 2000. Editors: F. Amoroso, U. Zannier. XI,353 pages. 2003.

Vol. 1820: F. Hiai, H. Kosaki, Means of Hilbert Space Operators. VIII, 148 pages. 2003.

Vol. 1821: S. Teufel, Adiabatic Perturbation Theory in Quantum Dynamics. VI, 236 pages. 2003.

Vol. 1822: S.-N. Chow, R. Conti, R. Johnson, J. Mallet-Paret, R. Nussbaum, Dynamical Systems. Cetraro, Italy 2000. Editors: J. W. Macki, P. Zecca. XIII, 353 pages. 2003.

Vol. 1823: A. M. Anile, W. Allegretto, C. Ringhofer, Mathematical Problems in Semiconductor Physics. Cetraro, Italy 1998. Editor: A. M. Anile. X, 143 pages. 2003.

Vol. 1824: J. A. Navarro González, J. B. Sancho de Salas, C^{∞} - Differentiable Spaces. XIII, 188 pages. 2003.

Vol. 1825: J. H. Bramble, A. Cohen, W. Dahmen, Multiscale Problems and Methods in Numerical Simulations, Martina Franca, Italy 2001. Editor: C. Canuto. XIII, 163 pages. 2003.

Vol. 1826: K. Dohmen, Improved Bonferroni Inequalities via Abstract Tubes. Inequalities and Identities of Inclusion-Exclusion Type. VIII, 113 pages, 2003.

Vol. 1827: K. M. Pilgrim, Combinations of Complex Dynamical Systems. IX, 118 pages, 2003.

Vol. 1828: D. J. Green, Gröbner Bases and the Computation of Group Cohomology. XII, 138 pages, 2003.

Vol. 1829: E. Altman, B. Gaujal, A. Hordijk, Discrete-Event Control of Stochastic Networks: Multimodularity and Regularity. XIV, 313 pages, 2003.

Vol. 1830: M. I. Gil', Operator Functions and Localization of Spectra. XIV, 256 pages, 2003.

Vol. 1831: A. Connes, J. Cuntz, E. Guentner, N. Higson, J. E. Kaminker, Noncommutative Geometry, Martina Franca, Italy 2002. Editors: S. Doplicher, L. Longo. XIV, 349 pages. 2004.

Vol. 1832: J. Azéma, M. Émery, M. Ledoux, M. Yor (Eds.), Séminaire de Probabilités XXXVII. XIV, 448 pages. 2003.

Vol. 1833: D.-Q. Jiang, M. Qian, M.-P. Qian, Mathematical Theory of Nonequilibrium Steady States. On the Frontier of Probability and Dynamical Systems. IX, 280 pages, 2004.

Vol. 1834: Yo. Yomdin, G. Comte, Tame Geometry with Application in Smooth Analysis. VIII, 186 pages, 2004.

Vol. 1835: O.T. Izhboldin, B. Kahn, N.A. Karpenko, A. Vishik, Geometric Methods in the Algebraic Theory of Quadratic Forms. Summer School, Lens, 2000. Editor: J.-P. Tignol. XIV, 190 pages. 2004.

Vol. 1836: C. Năstăsescu, F. Van Oystaeyen, Methods of Graded Rings. XIII, 304 pages, 2004.

Vol. 1837: S. Tavaré, O. Zeitouni, Lectures on Probability Theory and Statistics. Ecole d'Eté de Probabilités de Saint-Flour XXXI-2001. Editor: J. Picard. VII, 315 pages. 2004.

Vol. 1838: A.J. Ganesh, N.W. O'Connell, D.J. Wischik, Big Queues. XII, 254 pages, 2004.

Vol. 1839: R. Gohm, Noncommutative Stationary Processes. VIII, 170 pages, 2004.

Vol. 1840: B. Tsirelson, W. Werner, Lectures on Probability Theory and Statistics. Ecole d'Eté de Probabilités de Saint-Flour XXXII-2002. Editor: J. Picard. VII, 200 pages. 2004.

Vol. 1841: W. Reichel, Uniqueness Theorems for Variational Problems by the Method of Transformation Groups, XIII, 152 pages. 2004.

Vol. 1842: T. Johnsen, A.L. Knutsen, K3 Projective Models in Scrolls. VIII, 164 pages. 2004

Vol. 1843: B. Jefferies, Spectral Properties of Noncommuting Operators, VIII, 184 pages. 2004

Vol. 1844: K.F. Siburg, The Principle of Least Action in Geometry and Dynamics, XII, 128 pages. 2004.

Vol. 1845: Min Ho Lee, Mixed Automorphic Forms, Torus Bundles, and Jacobi Forms, X, 239 pages. 2004

Vol. 1846: h. Ammari, H. Kang, Reconstruction of Small Inhomogeneities from Boundary Measurements, IX, 238 pages. 2004

Vol. 1848: M. Abate, J. E. Fornaess, X. Huang, J. P. Rosay, A. Tumanov, Real Methods in Complex and CR Geometry, Martina Franca, Italy 2002. Editors: D. Zaitsev, G. Zampieri. IX, 219 pages. 2004.

Vol. 1849: Martin L. Brown, Heegner Modules and Elliptic Curves, X, 517 pages. 2004

Vol. 1850: V. D. Milman, G. Schechtman (Eds.), Geometric Aspects of Functional Analysis. Israel Seminar 2002-2003. X, 301 pages. 2004.

Vol. 1851: O. Catoni, Statistical Learning Theory and Stochastic Optimization, VIII, 273 pages. 2004.

Vol. 1852: A.S. Kechris, B.D. Miller, Topics in Orbit Equivalence, X, 134 pages. 2004.

Vol. 1853: Ch. Favre, M. Jonsson, The Valuative Tree, XV, 234 pages. 2004.

Vol. 1854: O. Saeki, Topology of Singular Fibers of Differential Maps, X, 145 pages. 2004.

Vol. 1855: G. Da Prato, P.C. Kunstmann, I. Lasiecka, A. Lunardi, R. Schnaubelt, L. Weis, Functional Analytic Methods for Evolution Equations, Levico Terme, Italy 2001. Editors: M. Iannelli, R. Nagel, S. Piazzera. VIII, 472 pages. 2004.

Recent Reprints and New Editions

Vol. 1200: V. D. Milman, G. Schechtman, Asymptotic Theory of Finite Dimensional Normed Spaces. 1986. – Corrected Second Printing. X, 156 pages. 2001.

Vol. 1471: M. Courtieu, A.A. Panchishkin, Non-Archimedean L-Functions and Arithmetical Siegel Modular Forms. – Second Edition. VII. 196 pages. 2003.

Vol. 1618: G. Pisier, Similarity Problems and Completely Bounded Maps. 1995 – Second, Expanded Edition VII, 198 pages. 2001.

Vol. 1629: J.D. Moore, Lectures on Seiberg-Witten Invariants. 1997 – Second Edition. VIII, 121 pages. 2001.

Vol. 1638: P. Vanhaecke, Integrable Systems in the realm of Algebraic Geometry. 1996 – Second Edition. X, 256 pages. 2001.

Vol. 1702: J. Ma, J. Yong, Forward-Backward Stochastic Differential Equations and Their Applications. 1999. – Corrected Second Printing. XIII, 270 pages. 2000.